Hormones and Metabolism in Insect Stress

Editors

Jelisaveta Ivanović
Scientific Advisor
Department of Insect Physiology and Biochemistry
Institute for Biological Research
Beograd, Yugoslavia

Miroslava Janković-Hladni
Scientific Advisor
Department of Insect Physiology and Biochemistry
Institute for Biological Research
Beograd, Yugoslavia

CRC Press
Taylor & Francis Group
Boca Raton London New York

CRC Press is an imprint of the
Taylor & Francis Group, an **informa** business

First published 1991 by CRC Press
Taylor & Francis Group
6000 Broken Sound Parkway NW, Suite 300
Boca Raton, FL 33487-2742

Reissued 2018 by CRC Press

Library of Congress Cataloging-in-Publication Data

Hormones and metabolism in insect stress / editors Jelisaveta
    Ivanović, Miroslava Janković-Hladni.
        p.   cm.
    Includes bibliographical references and index.
    ISBN 0-8493-6224-5
    1. Insects--Effect of stress on. 2. Insect hormones. 3. Insects-
-Metabolism. I. Ivanović, Jelisaveta, 1928-   . II. Janković
-Hladni, Miroslava, 1932-
    QL495.H64   1991
    595.7'01--dc20

A Library of Congress record exists under LC control number: 90047154

ISBN 13: 978-1-315-89413-3 (hbk)
ISBN 13: 978-1-351-07323-3 (ebk)

Visit the Taylor & Francis Web site at http://www.taylorandfrancis.com and the
CRC Press Web site at http://www.crcpress.com

PREFACE

Stress is a complex and general biological phenomenon, the initial link in the process of adaptation of populations to unfavorable environmental factors. Stress occurs in all living systems, from *Prokaryota* to *Eukaryota* including man, in a response to adverse factors (stressors). This phenomenon evolved very early in the evolution of living beings, probably with the evolvement of living matter.

The studies concerning the process of stress attract even greater attention of researchers in view of the fact that in the natural environment, in addition to natural stressors (e.g., temperature, humidity, lack of food), new stressors are increasingly included—pollutants, which endanger all living beings to a lesser or greater extent.

We maintained that it would be of theoretical and practical significance to present hitherto obtained results concerning the stress in insects. An attempt was made to recognize, at least in part, the complex phenomenon of stress in this most numerous, by species, and most various class, by life strategies, of animals. In addition, we bore in mind the economic aspect of the issue since some harmful species of insects, whose populations overcrowd, may do great damage to crops and forests. The authors believe that this book will prompt the interest of researchers dealing with both fundamental and practical studies. It is addressed primarily to entomologists of diverse profiles: ecologists, ecophysiologists, physiologists, applied entomologists and others. However, it is sincerely hoped that it will prove of value to all who work on stress using different model animals and dealing with different aspects. It should be noted that investigations of stress in insects are still in the pioneering stage, and thus we are keenly aware that generalizations presented in this book, based on the known facts, are also in the same stage. Finally, we realize that the book may still contain many shortcomings, and probably errors, but if its readers were prompted to exchange of ideas, discussions, critical evaluations and further research, we should not be sorry for the efforts made.

Before extending our thanks to all those who contributed directly to the preparation of this book, we feel necessary to express our deep gratitude to our late professors, Academician Sinisa Stanković, ecologist, and to Academician Ivan Giaja, physiologist, who influenced our way of thinking, aroused our enthusiasm for scientific work, and stimulated our interest for the research of general biological phenomena.

We greatly appreciate the readiness of our colleagues from the Soviet Union and Poland, who are involved in the elucidation of the mechanism of stress in insects, to provide original and significant contributions (S. I. Chernysh, Chap. 4; B. Cymborowski, Chap. 5; I. Yu Rauschenbach, Chap. 6) making this book more complete, particularly from the aspect of the hormonal role in the process of stress in insects.

It is also a pleasure to thank all of our research fellows, members of the Laboratory of Insect Physiology and Biochemistry, Institute for Biological Research "Sinisa Stanković": Dr. V. Stanić, Dr. M. Milanović, V. Nenadović, M. Božidarac, M. Frušić, V. Perić and S. Djordjević for the efforts they made and enthusiasm they showed in conducting our joint studies on stress in phytophagous insects.

Finally, we wish to express our thanks and appreciation to translators for translating some parts of this book: Dr. V. Stanić (Chapters 3 and 6) and Mrs. D. Filipović (Chapters 1, 2, 8 and 9).

J. P. Ivanović
M. I. Janković-Hladni
Belgrade
April 1990

THE EDITORS

Jelisaveta Ivanović, D.Sc., scientific advisor, is Chief, Laboratory of Insect Physiology and Biochemistry in the Institute for Biological Research "Siniša Stanković", Belgrade, Yugoslavia.

Dr. Ivanović obtained her training in biology at the Faculty of Natural Sciences, Belgrade University, earning B.S. and D.Sc. degrees in 1953 and 1963, respectively.

She served as Assistant Professor from 1956 to 1966 at the Faculty of Natural Sciences (Zoology), Belgrade University. Since 1966 she has been employed in the Institute for Biological Research "Siniša Stanković", holding the position of Chairman, Department of Physiological Ecology of Animals until 1973. In 1973, this Department was divided in two laboratories; one of them, the Laboratory of Insect Physiology and Biochemistry, was headed by Dr. Ivanović. Dr. Ivanović was engaged as Associate Professor for the postgraduate course entitled, "Nervous System in Invertebrates" at Belgrade University, Centre for Multidisciplinary Studies in 1973.

Dr. Ivanović is a member of the Serbian Biological Society; the Union of Yugoslav Physiologists; the Union of Yugoslav Biochemical Societies; and the Yugoslav Entomological Society. Dr. Ivanović was President of the Serbian Entomological Society from 1978 to 1981. During this period she managed to bring together the scientists of different profiles from fundamental to applied entomological research. Dr. Ivanović has been a member of the European Society for Comparative Endocrinology since 1972.

Dr. Ivanović has been recipient of many research grants: from the Serbian Republic Scientific Committee; from the U.S.D.A., Agricultural Research Service, for the project entitled "Endocrinological Studies of Phytophagous Insects" (1985—1988); and from Academy of Sciences U.S.S.R., Institute of Cytology and Genetics (1990).

Dr. Ivanović is the author of approximately 80 scientific papers in international and Yugoslav scientific journals, concerning mainly stress and adaptation syndrome in phytophagous insects, which is also her current research interest.

Miroslava Janković-Hladni, D.Sc., is scientific adviser, Laboratory of Insect Physiology and Biochemistry in the Institute for Biological Research "Siniša, Stanković", Belgrade, Yugoslavia.

Dr. Janković-Hladni graduated from Belgrade University with a B.S. degree in Biology. She obtained her M.S. degree in 1962, and D.Sc. degree in 1971 from the same University. Dr. Janković-Hladni was recipient of two scholarships: Hebrew University, Faculty of Agriculture, Rehovot, Israel (10 months) in 1960/61; and College de France, Paris (2 months) in 1965.

Dr. Janković-Hladni is a member of the Serbian Biological Society; the Union of Yugoslav Physiologists; the Union of Biochemical Societies; and the Yugoslav Entomological Society. During 1978—1981 she was scientific secretary of the Serbian Entomological Society in the peak of its activity. Dr. Janković-Hladni has been a member of the European Society for Comparative Endocrinology since 1969.

Dr. Janković-Hladni was recipient of the following grants: U.S.D.A., Agricultural Research Service for the project entitled, "Endocrinological Studies of Phytophagous Insects" (1985—1988); Academy of Sciences U.S.S.R., Institute for Cytology and Genetics (1990). She participated in many projects supported by the Serbian Republic Scientific Committee.

Dr. Janković-Hladni is the author of more than 60 scientific papers published in international and Yugoslav journals. Her current major research interest refers the role of hormones and enzymes in the development and stress in phytophagous insects.

CONTRIBUTORS

Sergey I. Chernysh, Ph.D.
Senior Research Worker
Biological Institute
Leningrad University
Leningrad, U.S.S.R.

Bronisław Cymborowski, Ph.D.
Professor
Department of Invertebrate Physiology
Warsaw University
Warsaw, Poland

Jelisaveta P. Ivanović, D.Sc.
Scientific Advisor
Department of Insect Physiology and
  Biochemistry
Institute for Biological Research
Belgrade, Yugoslavia

Miroslava Janković-Hladni, D.Sc.
Scientific Advisor
Department of Insect Physiology and
  Biochemistry
Institute for Biological Research
Belgrade, Yugoslavia

Inga Yu. Rauschenbach, D.Sc.
Senior Research Worker
Institute of Cytology and Genetics
Siberian Department of the U.S.S.R.
Academy of Science
Novosibirsk, U.S.S.R.

Valentin B. Sapunov, Ph.D.
Department of Control of Medical and
  Biological Systems
Leningrad State University
Leningrad, U.S.S.R.

TABLE OF CONTENTS

Chapter 1

INTRODUCTION

Jelisaveta P. Ivanović and Miroslava I. Janković-Hladni

In the early 20th century Cannon[1] recognized that, under stressory conditions, peristalsis in the dog stomach is intensified, which was subsequently related to the sympathetic nervous system, i.e., to the substance synthesized by this system. Upon being semipurified it was termed sympathin.

Somewhat later, distinguished Canadian physiologist Selye,[2] who dedicated all his life to the study of the phenomenon of stress in homeotherms, put forward the hypothesis on the general adaptation syndrome (GAS) defining stress as a nonspecific response of the organism to any demand upon it. In the last decade, thanks to explosive development of neurophysiology and endocrinology, it became clear that in homeotherms the hypophysis and adrenal gland cortex play the master role in the process of stress.[3] It was also speculated that quantitative rather than qualitative criteria were more important for the evaluation of stress.[4]

However, the advancements in comparative physiology and endocrinology as well as decrement of fish found in the rivers and seas prompted scientists to study the process of stress in lower vertebrates (fishes). The results obtained reveal the existence of stress in fishes. They also indicate that stress is caused not only by chemical substances (pollutants) but also by the fishes' limited swimming range (construction of dams) and behavioral factors (migration). More close endocrinological examination of the mechanisms of stress in fishes showed that protective-adaptive reactions of the organisms are regulated chiefly by three morphological-functional systems: hypothalamo-anteriorhypophyseal, hypothalamo-meta-hypophyseal and hypothalamo-posthypophyseal.[5]

It is interesting to point out that investigations concerning the control of pest insects by chemical substances played the key role in starting the studies on the response of insects to the action of these substances. Sternburg[6] must be given great credit for the advancement of these investigations. Namely, even at that time (1963) he suggested the significance of autointoxication (endogenous factors) influenced by physical and chemical factors. He reported that intensive stress induces autointoxication which in turn leads to paralysis and even death of insects.

After Sternburg's paper sporadic papers on the response of insects to stress action were published. The majority of these authors considers that this response is not identical but similar to the process of stress in higher vertebrates; only a small number of authors maintains that this phenomenon in insects is identical to the process of stress in vertebrates.[7-9]

Cook and Holman[9] define stress in insects as the response of an organism to meet all adaptive demands made from the environment. The response of an insect is dependent on the complex and integrated release of chemical messengers.

As soon as the investigations on stress, as a biological phenomenon, were undertaken the question of whether these responses are specific or nonspecific was posed. Many scientists, dealing with the phenomenon of stress in mammals, maintain that this response is specific.[10,11] Some authors argue that the stress is a nonspecific response.[12] Others are of the opinion that some responses in homeotherms are specific and some nonspecific.[13,14] Extensive studies concerning stress in the wide range of species revealed similarities in some responses to stressor, which is of particular significance from the evolutive aspect.

A paradigm for this similarity is the synthesis of heat shock proteins, which was observed in a variety of organisms spanning Protozoa,[14] plants,[15] insects,[16] and homeotherms.[17] To support the generality of the response it should be noted that this group of specific proteins is synthesized in the nucleus not only in response to high temperatures but also under the effect of chemical and other environmental factors.[18] The function of heat shock proteins (hsp) is not entirely evaluated. However, many authors, including Petersen and Mitchell[18] for example, maintain that these specific proteins improve the resistance of organisms in responding to stressors. Assuming that hsp behavior is analogous to that of microsomal oxidases it seems probable that their synthesis, caused by other stressors, is secondary. Namely, it is well known that the activity of microsomal oxidases, synthesized in the tissues of phytophagous insects in response to some chemical substances in plants, is increased under the influence of insecticides and temperature. Bearing in mind the fact that the evolution of living matter is associated with the regimen of relatively high temperatures as well as the fact that hsp are present also in microorganisms, it may be assumed that hsp are evolutively very old.

The similarity in the response between lower vertebrates (amphibians and fishes) and invertebrates (insects, centipedes, mollusks) to the effect of subzero temperatures was observed. Organisms are vulnerable to subzero temperatures because they cause ice formation not only in the extracellular fluid but also in cells. Both in lower vertebrates and in insects one of the protective responses to subzero temperatures is the elevation of blood glucose level which is most frequently controlled by hormones. In amphibians, in the first phase of stress, the elevation of glucose level is caused by catecholamines and subsequently by glycogen. In insects during the first phase of low temperature stress the increase in trehaloze level is influenced by octopamine and subsequently by hypertrehalosemic hormone and adipokinetic hormone (AKH). Storey and Storey[19] maintain that elevation of glucose level, under the effect of low temperatures, has adaptive significance because it results in the stabilization of subcellular structures and proteins as well as in the suppression of metabolism. Namely, Hochachka and Guppy[20] demonstrated in winter frogs that the acclimation process to low temperatures is primarily characterized by the suppression of metabolism, i.e., by the transition into hypometabolic state in which the organism is far less susceptible to anoxia and ischaemia which are induced by low temperature. This phenomenon (suppression of metabolism) is one of the characteristics of diapause,[21] which reflects the adaptation of insects to unfavorable environmental conditions and during which nonspecific resistance of the organism to the effect of different factors is markedly pronounced.[22]

More recently published cryophysiological papers concerning the study of the action of low temperatures at the molecular level indicate that these temperatures provoke a number of general reactions such as changing of catalytic efficiency channels, pumps and metabolic enzymes as well as modulating of the membrane and metabolic catalysts.

Bearing in mind all these facts, our current opinion is that stress in insects is the state of the organism caused by one factor or by a combination of factors, affecting the survival limits in different ways. Stress reflects an interrelated complex of different physiological, biochemical, and behavioral responses which are manifested in hierarchical order at different levels of biological organization. As far as the specificity is concerned, these responses are within the range whose limits outline both general nonspecific changes and specific changes characteristic of the response of the specific tissue or species.

Almost 60 years ago Cannon expressed his opinion by saying: "The effect of emotional stress and the mode of relieving it offer many opportunities for useful studies. I heartily commend it to you." His commendation is still valid since though the mechanism of stress in homeotherms has been extensively studied ever since it is not thoroughly recognized nor was the general model proposed. Data on stress in insects, on which relevant investigations were initiated relatively recently, are less numerous. If this fact is taken into account along with the great plasticity of insects as a group, being reflected in the great radiation of its species, it is obvious that establishment of such a model system for stress in insects would not be possible in the near future. The lack of such a system will also slow down evolutive recognition of the mechanism of stress and definition of stress as a general phemonenon of the living matter.

In this book an attempt has been made to present data from still pioneering papers on stress in insects. We are aware that for some objective and subjective reasons it must have some shortcomings. Nevertheless, we hope to be forgiven because, as wittingly remarked by Napoleon Bonaparte's marshals, "It is difficult to be one's own ancestor." However, if this book succeeds only in prompting further progress of the investigations on stress in insects, we should believe that it greatly achieved its objective.

REFERENCES

1. **Cannon, W. B.,** Bodily changes in pain, hunger, fear and rage, in *Account of Recent Researches into the Function of Emotional Excitement,* Appleton Ged., New York, 1929.
2. **Selye, H.,** Syndrome produced by diverse nocuous agent, *Nature,* 138, 32, 1936.
3. **Axelrod, J.,** Catecholamines and stress: a brief informal history, in *Catecholamines and Stress: Recent Advances,* Usdin, E., Kventansky, R., and Kopin, I. J., Eds., Elsevier, Amsterdam, 1980, 3.
4. **Kozlowski, S.,** Panel discussion on stress theory, in *Catecholamines and Stress: Recent Advances,* Usdin, E., Kventansky, R., and Kopin, I. J., Eds., Elsevier, Amsterdam, 1980, 583.
5. **Polenov, A.,** Obscij princip gipotalamiceskoj nejroendokrinoj reguljacii zascitno priposobiteljnih reakcij organizma, in *Endokrinaja Sistema Organizma i Toksiceskie Faktori Vneshnej Sredi,* Akademija nauk SSSR, 1980, 272.
6. **Sternburg, J.,** Autointoxication and some stress phenomena, *Annu. Rev. Entomol.,* 8, 19, 1963.
7. **Ivanović, J., Janković-Hladni, M., Stanić, V., Milanović, M., and Nenadović, V.,** Possible role of neurohormones in the process of acclimitization and acclimation in *Morimus funereus* larvae (Insecta). I. Changes in the neuroendocrine system and target organs (midgut, haemolymph) during the annual cycle, *Comp. Biochem. Physiol.,* 63A, 95, 1979.
8. **Janković-Hladni, M., Ivanović, J., Stanić, V., Milanović, M., and Božidarac, M.,** Possible role of neurohormones in the process of acclimation and stress in *Tenebrio molitor* and *Morimus funereus.* I. Changes in the activity of digestive enzymes, *Comp. Biochem. Physiol.,* 67A, 477, 1980.
9. **Cook, B. J. and Holman, M. G.,** Peptides and kinins, in *Comprehensive Insect Physiology, Biochemistry and Pharmacology,* Vol. 11, Kerkut, G. A., and Gilbert, L. J., Eds., Pergamon Press, Oxford, 1985, 532.
10. **Ganong, W. F.,** Participation of brain monoamines in the regulation of neuroendocrine activity under stress, in *Catecholamines and Stress: Recent Advances,* Usdin, E., Kventansky, R., and Kopin, I. J., Eds., Elsevier, Amsterdam, 1980, 115.
11. **Vigas, M.,** Contribution to the understanding of the stress concept, in *Catecholamines and Stress: Recent Advances,* Usdin, E., Kventansky, R., and Kopin, I. J., Eds., Elsevier, Amsterdam, 1980, 583.
12. **LeBlanc, J.,** Stress and interstress adaptation, *Fed. Proc.,* 28, 996, 1969.
13. **Petrovic, V.,** Prilog poznavanju neuroendokrine regulacije termo adaptacionih i ciklicnih procesa, *Extrait du Glas de l'Academie Serbe des Sciences et des Arts,* 51, 1987, 139.
14. **Kalinina, L., Khrebtukova, I., Podgornaya, O., Wasik, A., and Sikora, J.,** Heat shock proteins in amoeba, *Eur. J. Protistol.,* 24, 64, 1988.
15. **McAllister, L. and Finkelstein, B.,** Heat shock proteins and thermal resistance in yeast, *Biochem. Biophys. Res. Commun.,* 93, 819, 1980.
16. **Tissiers, A., Herschel, K. M., and Tracy, U.,** Protein synthesis in salivary glands of *Drosophila melanogaster:* relation to chromosome puffs, *J. Mol. Biol.,* 84, 389, 1974.
17. **Mitchell, H. U., Moller, G., Peterson, N. S., and Lipps-Sarmiento, L.,** Specific protection from phenocopy induction by heat shock, *Dev. Genet.,* 1, 181, 1979.
18. **Petersen, N. and Mitchell, H.,** Heat shock proteins, in *Comprehensive Insect Physiology, Biochemistry and Pharmacology,* Kerkut, G. A. and Gilbert, L. I., Eds., Pergamon Press, Oxford, 1985, 347.
19. **Storey, K. and Storey, J.,** Freeze tolerance in animals, *Physiol. Rev.,* 68, 28, 1988.
20. **Hochachka, P. W. and Guppy, M.,** *Metabolic Arrest and the Control of Biological Time,* Harvard University Press, Cambridge, MA, 1987.
21. **Behrens, W.,** Environmental aspects of insect dormancy, in *Environmental Physiology and Biochemistry of Insects,* Hoffman, K., Ed., Springer-Verlag, Berlin, 1985, 67.
22. **Denlinger, D., Giebultowich, J., and Adedokun, T.,** Insect diapause: dynamics of hormone sensitivity and vulnerability to environmental stress, in *Endocrinological Frontiers in Physiological Ecology of Insects,* Sehnal, F., Zabza, A., and Denlinger, D., Eds., Wroclaw Technical University Press, Wroclaw, Poland, 1988, 309.

Chapter 2

HORMONES AND METABOLISM IN INSECT STRESS
(Historical Survey)

Miroslava I. Janković-Hladni

TABLE OF CONTENTS

I. INTRODUCTION

It is not possible to speak about the history of stress investigations in insects without mentioning the world reknowned, distinguished Canadian scientist, Hans Selye,[1] who discovered this phenomenon as early as 1936. Selye[2] also put forward the concept of stress as a general adaptive syndrome and over 50 years provided voluminous literature on the adaptation of living systems contributing immensely to this field.

In view of the fact that authors of this book attached extensive reference lists for each chapter, we shall outline data obtained so far on basic characteristics and links of the process of stress in insects. Therefore, one should bear in mind the number of insect species (each day new species are discovered) as well as the fact that out of all living beings, insects occupy the greatest number of ecological niches, even those most uncommon (textile, iron, wax, etc.), which compounds the difficulty of discovery of general mechanisms of the process of stress in this class of animals.

Editors of this book consider stress, i.e., response to the effect of stressors, as a very complex and dynamic process which is the prerequisite for the adaptation and evolution of all living beings and hence also of insects.

The studies of the process of stress in insects started in the early 1950s and are closely related to the studies of insecticides (effects, mechanisms). As early as 1952, Sternburg and Kearns[3] established that in the hemolymph of DDT-treated cockroaches toxins were synthesized which were neither DDT nor its metabolites. Already Wigglesworth[4] had visionary ideas on the role of the neuroendocrine system in the process of stress by speculating: ''The corpus cardiacum might be involved in the secretion not only of growth hormones but of a pharmacologically active principle or principles with those produced by the pituitary or suprarenal glands of vertebrates.''

Subsequently, in the course of some 10 years the studies of stress involved other stressors as well (mechanical, salinity stress, electrical stimulation, etc.) which enabled Sternburg[5] to sum up the results and hypothesize on the autointoxication. This author is of the opinion that, regardless of the kind of stressor, stress of great intensity stimulates the insect to release its own pharmacologically active substances which cause the paralysis or even death of stress-treated insects due to autointoxication.

After Sternburg's review article was published, in which the author summed up the results on stress in insects obtained up until 1962, interest in these studies was on the wane. It was not until the late 1970s and in the 1980s that renewed interest for these studies resulted in articles dealing with general aspects of stress in insects. Thus, Ivanović et al.,[6,7] and Chernysh[8] reported on the dominant role of the neuroendocrine system in the process of stress in insects. Cook and Holman,[9] however, maintain that in addition to hormones some other substances from the central nervous system (e.g., neurotransmitters) may also play a significant role in this process.

Whereas stress in insects was earlier considered solely as a pathological state, the prevailing opinion today is that stress is the link in the process of adaptation, including states ranging from metabolic hyperactivity to the pathological state leading to the insect's death.

II. BIOLOGICALLY AND PHARMACOLOGICALLY ACTIVE SUBSTANCES

An intriguing question on the origin and nature of substances involved in stress was posed. On the basis of experimental data and findings obtained by following up insect behavior, under the effect of stressors, it was concluded that these substances should be sought in the neuroendocrine and nervous systems. Parallel investigations on both these systems were carried out but we shall summarize only the results regarding the neuroendocrine system, starting from pioneering research to date.

A. THE RESPONSE AT THE LEVEL OF NEUROENDOCRINE SYSTEM
1. Neurosecretion

Classical general concept on neurosecretion was put forward by Scharrer and Scharrer[10] in 1944 and was concerned with invertebrates and vertebrates. Both vertebrates and invertebrates are characterized by the presence of specialized neurons which, in the perikaryon, synthesize neurosecretory material which, in insects, is transported along the axon to the corpora cardiaca (CC) from the protocerebrum and to other neurohemal organs from the other ganglia of the nervous chain. From the neurohemal organs the neurohormones are released into the hemolymph by exocytosis.

As early as 1917, Kopec[11] identified the nervous system of insects as a source of endocrine agents. He reported that brain regulates molting and metamorphosis in gypsy moth *Lymantria dispar* larvae. It was the first demonstration of endocrine activity in the nervous system in all the animal kingdom.

B. Scharrer, over the course of 50 years, conducted the pioneering research concerning neurosecretion in invertebrates, including insects, stimulating many studies in this field.[12,13]

The concept of neurosecretion is today substantially extended to include, in addition to classical neurosecretory cells, i.e., neurosecretory neurons which produce neurohormones, aminergic neurons which produce biogenic amines. Of particular interest for insects is octopamine which, in addition to playing the role of a transmitter, also plays the role of neurohormone. On the other hand, neurosecretory cells may send their product to the effector along the axon and not through the hemolymph. Such is the case with, e.g., proctolin, a pentapeptide, which is a myotropic substance with a potent activity on a number of invertebrate visceral, skeletal, and heart muscle preparations.

In this short historical outline there follow hereunder examples of the effect of different stressors on: (a) neurosecretory cells, (b) neurohormones, (c) paralyzing agents, and (d) biogenic amines.

a. Neurosecretory Cells

Presently, the most numerous data are those on the changes, caused by different stressors, in the activity of neurosecretory cells (synthesis, release), which are indirectly indicative of changes in the hemolymph neurohormones. Immunocytochemical studies on neurosecretory cells by means of specific antibodies will contribute to determining more precisely the localization of some hormones as well as the site of synthesis and release from certain cells.

Hodgson and Geldiay[14] were among the first to demonstrate that under the effect of physical stressor in the two species of cockroaches, i.e., in *Blaberus cranifer* (enforced activity) and in *Periplaneta americana* (electrical stimulation) the quantity of neurosecretory material (NSM) in the CC of both species was decreased. These authors suggested that NSM release from CC might be a normal response of these insects to the effect of stressors, expressed at the level of the neurosecretory system. The same authors indicated the possible analogy with the release of epinephrine (adrenaline) in mammals under stressful conditions. Colhoun[15] reported that in DDT treated *P. americana* the quantity of NSM in CC also diminishes. Davey[16] found in *P. americana*, which is the favorite model animal for these investigations, the release of cardiac accelerator from CC during enforced activity. Kater[17] showed in the same insect that electrical stimulation induced the release of cardiac accelerator, a proteinaceous heat stable substance, from CC. Numerous investigations concerning the effects of insecticides and other stressors on medial neurosecretory cells (MNC) as well as on other neurosecretory cells in the nervous chain (e.g., ventromedial neurosectory cells (VMNC) in the subesophageal ganglion) showed that various stressors of different intensity cause significant changes in the synthesis and release of neurosecretory material. Besides, histochemical changes are likely to occur.

We shall specify only some findings. Vojtenko[18] observed that in *Malacosoma neustria, Portherita dispar,* and *Leucoma salicis* in the brain MNC 24 h after the application of sublethal doses of phosphoamide or DDT the synthesis and release of NSM are intensified. These doses stimulated the neurosecretion without causing pathological changes. Lethal doses led to destructive changes in the neurosecretory and nervous cells. Similar changes in MNC in *Tenebrio molitor* were found by Janković-Hladni and Grozdanović[19] during acclimation of larvae to 13 and 31°C, whereby lower temperature stimulates NSM synthesis and release and higher temperature leads to the decrease in the NSM synthesis and NSM accumulation in the perikaryons. Ivanović et al.,[20,21] showed in larvae of the cerambycid beetle *Morimus funereus*, exposed to temperatures of 0, 8, and 23°C, that lower temperature (0°C) induces the increased synthesis in VMNC of the subesophageal ganglion (SG); temperature of 23°C provokes, after 15 days of acclimation, the decreased synthesis of NSM in VMNC of SG, and decreased release and possible change in histochemical properties of NSM. These and subsequent investigations showed that the response in different insect species at the level of neurosecretory cells depends on the intensity of the stressor and the duration of the exposure.[22,23] Ivanović et al.,[24] by studying comparative changes at the level of the proteocerebral neurosecretory cells as well as at the level of other biochemical parameters during the annual cycle in larval cerambycid beetle *M. funereus*, observed appreciable changes in the activity of lateral and medial neurosecretory cells. Owing to the results obtained they hypothesized on a dominant role of the protocerebral neurosecretory cells, i.e., of neurohormone(s) in the acclimatization of this insect.

It seems of interest to point out, though papers dealing with this field are sporadic, that neurosecretory cells respond selectively to various stressors under certain conditions. Norman[25] reported that lindane produces hypotrehalosemia in blowfly *Calliphora vicina* which is affected by MNC, whereas other insecticides, e.g., pyretroids, lead to hypertrehalosemia which is influenced by the intrinsic cells of CC. Selective response at the level of the protocerebral MNC and lateral neurosecretory cells (LNC) to the effect of temperature and other stressors was also established in larvae of two species of cerambycid beetle *Cerambyx cerdo* and *M. funereus*.[26,27] The authors suggested that differences in the membrane permeability between subtypes of the protocerebral neurosecretory cells result in the selective response of neurosecretory cells to the effect of different stressors.

b. Neurohormones

Neurohormones are the master regulators and in insects they affect a multiplicity of physiological processes including the secretion of JHs and ecdysteroids.

It is known, on the basis of the effect of crude extracts of neurohemal tissues, that neurohormones regulate a multitude of physiological processes during the development, reproduction, metabolism, and physiological homeostasis.[28] It is obvious that those physiological processes, which are controlled by neurohormones during development and homeostasis, are of primary importance for the process of stress. The neurohormones govern among others the release of lipids from the fat body-adipokinetic hormone;[29] release of carbohydrates from the fat body-hyperglycemic (hypertrehalosemic) hormone;[30] enhancement of carbohydrate uptake-hypoglycemic (hypotrehalosemic) hormone;[31] increment of water excretion-diuretic hormone;[32] suppression of water excretion-antidiuretic hormone;[33] stimulation of cytochrome synthesis to enhance fat body metabolic activity,[34] etc. The roles mentioned are ascribed to those neurohormones whose biochemical upset, in terms of either excessive release into hemolymph or in their deficiency provoked by stressory factors, may be of fundamental significance for the outcome of stress.

Results referring to changes in neurohormones during the acclimation process are scarce which is understandable if one bears in mind that until 1984 only two neuropeptides were isolated and identified: proctolin[35] and adipokinetic hormone (AKH),[36] due to insufficiently sensitive techniques for the isolation and identification of neuropeptides in insects. However, recent advances in sophisticated, ultrasensitive techniques such as reverse-phase high-performance liquid chromatography, fast atom bombardment mass spectrometry, and gas-phase aminoacid sequencing have solved many of the technical problems of peptide identification. Hitherto, 14 different neurohormones from the family AKH/RPC were isolated and sequenced in different insect species.[37] In addition to these, 20 other neurohormones were isolated.[38]

The study of adipokinetic hormones in *Locusta, Schistocerca,* and other insect species from biochemical and physiological aspects[39] opened new avenues for the study of stress.

Investigations conducted in the early 1970s on the mechanisms of the influence of insecticides on *Rhodnius prolixus,*[40] *Schistocerca gregaria,*[41] and *Periplaneta americana*[42] indicated that these stressors provoke excessive release of several neurohormones—diuretic, hyperglycemic, and adipokinetic—which results in the insect's death if doses are lethal.[43] Maddrell,[44] summarizing the previous results, concluded that the neuroendocrine system of insects is the target for insecticides. Somewhat earlier, indirect proof of the excessive release of neurohormones at the subcellular level was provided by Norman and Samaranayaka-Ramasamy[45] by examining CC in lindane-treated *Schistocerca gregaria.* The authors observed the presence of numerous exocytosis-omega figures, increased numbers of mitochondria and dense mitochondrial granules. They also suggested that lindan directly affects the intrinsic cell membrane making it permeable to calcium ions. Orchard and Osborne,[45a] on the grounds that some insecticides increase action potential in *Carausius morosus,* also suggested that insecticides influence the permeability of the neurosecretory cell membrane.

For more details on the role of neurohormones in the process of stress, the reader is referred to Section III of this chapter, and to Chapters 3 and 4.

c. Paralyzing Agents

In view of the fact that under the effect of different stressors (mechanical, electrical stimulation, insecticides) similar changes occurred in behavior (hyperactivity, incoordination of muscular movements, paralysis, prostration) it seemed likely that the nervous system releases some substances which produce change in the activity of insects and bring about their prostration or death. As early as in the 1950s Beament,[46] who explored the effect of mechanical stressors in the cockroach *P. americana,* concluded that insects of this species, after being under the effect of this stressor for some time, become paralyzed. The same author, on the basis of parabiotic experiments, by injecting the hemolymph of paralyzed individuals into nontreated animals produced their paralysis. Thus, he concluded that the main culprit for this paralysis is some factor which is transported via the hemolymph.

Somewhat later similar results were obtained by Colhoun[47] who treated the insects with chemical stressors, i.e., insecticides. By means of parabiotic experiments he also caused paralysis in nontreated insects. Interestingly, the presence of some other substances in the blood of DDT treated cockroaches has been previously shown.[3]

The question arose from where these pharmacologically active substances originate. Data of Sternburg et al.[48] demonstrated that if the hemolymph of previously DDT-treated cockroaches or that of electrically stimulated cockroaches is applied directly to the nervous chain of untreated animals its stimulation is provoked. These findings indicated the possibility that pharmacologically active substances are released from the nervous or neurosecretory system.

Subsequently, more detailed investigations of Cook et al.[49,50] on *P. americana* suggested the presence of a neuromuscular excitatory substance in the hemolymph, the concentration of which was doubled under the effect of mechanical stress. Chemical structure of this substance, i.e., of factor S was not determined but it was proved that this factor differs chemically, chromatographically, and biologically from acetylcholine, epinephrine, norepinephrine, histamine, γ-aminobutyric acid, and 5-hydroxytryptamine.

d. Biogenic Amines

The studies on the process of stress in insects are closely correlated with the progress of these studies in vertebrates, which seems understandable in view of the general character of this phenomenon and the role it plays in the evolution of the living systems. Findings that biogenic amines play a role in the process of stress in vertebrates undoubtedly raised the interest to study and identify these compounds in insects. A scientific quest was underway to answer the question of whether biogenic amines have, in this class of animals, a role similar to the one they have in the stress of vertebrates. The studies concerning biogenic amines and their role in the process of insect stress are a relatively new field; intensive studies started some 10 years ago. The results of these studies showed that octopamine is one of the most common biogenic amines. Its presence was discovered in the nervous and neuroendocrine systems[51] as well as in the hemolymph of many insect species such as *Locusta migratoria*,[52] *Schistocerca gregaria*,[53] and other insects.

In the early 1970s it was established that in the cockroach *P. americana*, various stressful factors elicit the increase in glucose[54] and trehalose contents[55] in the hemolymph. Downer[56] speculated that octopamine in the hemolymph of insects may function as a neurohormone controlling carbohydrate and lipid metabolism as the primary response of the insect to the effect of stressor. Subsequent investigations of Downer et al.[57] showed that in *P. americana* the hypertrehalosemic factor is influenced by octopamine suggesting that this factor is released from the octopaminergic neurone via nervi corpori cardiaci II (NCC II). It is assumed that biogenic amines are responsible for stressogenic mobilization of carbohydrates.

Octopamine in insects functions both as circulatory neurohormone and neuromodulator.[52] Davenport and Evans[59] suggested that it seems likely that octopamine may function in the insect hemolymph as a neurohormone controlling short-term lipid and carbohydrate metabolism, being released as a part of the arousal mechanism in response to stressors. The same authors in the same paper reported that in *S. gregaria* and in *P. americana* various stressors, i.e., physical (mechanical, thermal) and chemical (five types of insecticides) produce the increase in octopamine concentration. Davenport and Evans[59] concluded that the response of insects to stressors is a very complex process. They also argued that changes in octopamine concentration, influenced by stressors, reflect only one of the actions of numerous pharmacological compounds which are freely released under these conditions. Octopamine release under stressory conditions is ideally suited to fulfill a role in general response to stress in insects. They consider that octopamine functions as a neurohormone which elevates short energy supply to muscles and so mediates a flight or fight in insects. Orchard also maintains that octopamine may have a multiple role in insects, being neurotransmitter, neurohormone, and neuromodulator.[60]

Researchers were particularly tempted to study the role of octopamine in social stress in insects. The results obtained are interesting, though contradictory. The results of Fuzeau-Braesch[61] show that solitary forms of *L. migratoria* have higher octopamine concentration than gregarious forms. In another species of grasshoppers (*S. americana gregaria*) Morton and Evans[62] found no differences in octopamine concentration between solitary and gregarious phase. Kozánek et al.,[63] discovered that increased population density in *Mamestra brassicae* and *Nauphoeta cinerea* leads to the elevation of octopamine concentration in the ganglia of the central nervous system.

2. Ecdysteroids and Juvenile Hormones

Though there is a voluminous amount of information on these hormones from biochemical and physiological standpoints, data concerning the response of these hormones to stressors as well as their role in the process of stress are scarce.

It is well known that ecdysteroids and juvenile hormones (JHs) are growth and developmental hormones in insects and that their activity is governed by prothoracicotropic hormone (PTTH) and allatotropic hormone (ATTH). Normal development of the insect depends, in addition to other factors, on the ratio of the concentrations of these hormones in the hemolymph. It is also known that various stressors (injury, starvation, unfavorable temperatures, x-radiation, chemical and mechanical) may upset, directly or indirectly, hormonal balance, i.e., normal development. These disturbances may result in the insect's death if stressors are overwhelmingly intensive.

The first data ever on the effect of stressors on ecdysteroids and JH were reported in the early 1960s, and refer to injury as a stressory factor. Namely, Lüscher and Engelman[64] demonstrated that surgical damage causes in the cockroach *Leucophaea maderae* supernumerary moults which indirectly pointed to the increase in the JH concentration and delayed release of ecdysone. However, in diapausing *Saturnids* the injury produces elevation of ecdysone synthesis.[65] Subsequent studies revealed that diapausing insects, in the majority of cases, respond to stressors by increasing ecdysteroid concentrations, whereas nondiapausing insects respond otherwise. Actually, response to injury has a protective role. However, x-radiation injures insect endocrine cells selectively, irreversibly supressing ecdysteroid synthesis, as proposed by Piechowska.[66]

Temperature stress either below or above temperature optimum produces changes in hormonal balance.

In the mid 1970s Pipa[67] and Cymborowski and Bogus,[68] independently of each other, showed that exposure of last larval instar of *Galleria mellonella* to cold (0°C) causes supernumerary moults. As Cymborowski et al. had been dealing with the mechanism of the effect of cold stress for more than 10 years, their subsequent studies demonstrated that in cold-treated larvae JH titer, released by ATTH-activated corpora allata, is increased.[69,70] For details of these studies readers are referred to Chapter 5 of this book. Data on cytological investigations in larvae of *Cerambyx cerdo*, collected in November, show that corpora allata cells are activated during exposure to low temperature (10°C).[71]

Low temperature (0°C) leads also to delayed increase in ecdysteroid titer, as recently shown in the last larval instar of *G. mellonella*[72] and in larvae of *M. funereus*.[7] Ivanović et al.[7] correlate the increased ecdysteroid titer in *M. funereus* with the elevated activity of lateral neurosecretory cells (L_1), which are assumed to have a prothoracicotropic role as do protocerebral medial neurosecretory cells (A_1).

The above findings support Pipa's[67] suggestion, given about 15 years ago, that chilling stress has a stimulatory effect on both allatotropic and prothoracicotropic activities.

Extensive studies concerning cold hardiness in insects have been conducted intensively for more than 30 years. However, only recently researchers focused their attention on the study of the role of the neuroendocrine system in cold hardiness.

Horwath and Duman[73] reported that in *Dendroides canadensis* thermal hysteresis proteins (THPs) are controlled by J H. Baust et al.[74] suggested that the same hormone plays a role in the regulation of cryoprotectants (glycerol and sorbitol) as well as of ice nucleating agents (INA) in freeze-tolerant third larval instar of *Eurosta solidaginis*. However, quite opposite results are available. Thus, Nordin et al.,[75] on the basis of the findings on glycerol and on the role of the neuroendocrine system in the regulation of this cryoprotectant in the fifth larval instar of *Ostrinia nubilalis*, consider that the neuroendocrine system is not involved in the glycerol regulation. Recently Pullin and Bale,[76] studying the diapausing and nondiapausing pupae of *Pieris brassicae*, came to the conclusion that both environmental factors (e.g., photoperiod, temperature) and endogenous physiological factors (hormonal events) are involved in cold tolerance strategy.

Until more extensive, multidisciplinary work is done, through joint engagement of cryophysiologists and endocrinologists, we will not really know the role of the neuroendocrine system in cold hardiness.

It was found that starvation in *Manduca sexta* exerts an effect similar to that of cold stress, i.e., it produces supernumerary moult and increase in JH titer.[76,77] Viruses (nuclear polyhedrosis virus) also cause high JH titer in *Spodoptera litura*.[78]

Rauschenbach et al.[79-82] have been engaged, for more than 10 years, in the study of several stressors (high temperature, crowding, deficient diet) on metamorphosis in *Drosophila virilis* from a genetic-endocrinological standpoint. In addition to ecdysone and JH these studies also included the protocerebral medial neurosecretory cells, type A_1. Particular attention was paid to JH esterase, which is a JH degrading enzyme and is gene controlled. It is transformed from inactive to active form by an as yet unidentified activating factor from the brain. The authors maintain that JH esterase plays a key role for insect's survival, i.e., metamorphosis or death.

Mechanical stress, such as shaking of pupae *Antherea jamami* (*Lepidoptera:Saturnidae*) leads to premature completion of diapause in this insect[83] and electrical shock in *Bombyx mori* accelerates larval ecdysis[84] which is the implication in both cases of the increased ecdysone release under the effect of these stressors.

Chernysh[8,85] reported that chemical stress leads to the ecdysteroid increase, which was observed in caterpillars *Barathra brassicae* and *Calliphora vicina* when treated with formaldehyde vapors and methanol, respectively. This author hypothesizes on the protective role of ecdysone, maintaining that this hormone is one of the crucial links in the protective mechanisms of insects against stressors (for details readers are referred to Chapter 4). Ivanović et al.,[26] by applying temperature as stressor, investigated the basis of protective mechanism in *M. funereus* larvae. The results obtained imply that in protective mechanisms of this species against temperature stress both A_1 protocerebral medial neurosecretory cells, seemingly prothoracicotropic, and ecdysone play the major role.

Most recently, Bodnaryk[86] has suggested that in diapausing pupae of *Mamestra configurata* high temperature blocks development reversibly, which is restored under the effect of 20-hydroxyecdysone.

The aforementioned findings show that both hormones, under the effect of stressors, may have a protective role as well as that protective mechanisms are very complex; they include neurohormones (e.g., PTTH and ATTH) as well as hormone degrading enzymes out of which only JH-esterase has been studied from this aspect.

III. METABOLIC CHANGES

Although the studies on the process of stress in insects were carried out at random, still since the 1950s till today quite a lot of information has been accumulated concerning the effect of stressors on metabolic processes in insects. It has been proven or at least assumed that many metabolic changes occur under the control of hormones and biogenic amines, especially of octopamine. Metabolic changes in insects caused by stress have been extensively analyzed and summarized in Chapter 3 of this book.

A. CHANGES IN GENERAL METABOLISM

As early as the 1950s it was shown that under the influence of physical and chemical stressors significant changes in the metabolism of *Blatella*[87] occur. When this insect is DDT poisoned the oxygen consumption is increased by 3 to 4 times which coincides with the decrease in energy stores. Some 10 years later Heslop and Ray[88] in restrained and DDT poisoned cockroaches *P. americana* observed similar signs of paralysis. In both groups the oxygen consumption was increased at the start of the exposure to stressors and then, well before the exhaustion of endogenous reserves, it precipitously decreased. These authors suggested that insects, after entering the general syndrome of stress, cannot utilize energy reserves due to either biochemical upset or the upset of transport mechanism. The increase in oxygen consumption and, later, incoordination of movements as well as signs of paralysis were also observed during exposure of diapausing caterpillars *Ostrinia nubilalis* to upper lethal temperatures of 45°C.[89] These authors propose that this oxygen consumption is associated with the increased synthesis of neurosecretory material, i.e., of neurohormone(s) in the protocerebral (A$_1$) medial neurosecretory cells, which were studied later, under the same experimental conditions, in diapausing larvae of this species.[90]

B. CHANGES IN INTERMEDIARY METABOLISM

Appreciable changes at the level of intermediary metabolism, provoked by various stressors, were found during all phases of the process of stress. Changes in intermediary metabolism were studied best under cold stress. Cold stress, in respect to other stressors, has its specificities due to the action of subzero and freezing temperatures on insect species which inhabit polar and temperate regions on the earth, surviving temperatures as low as −70°C. These species, during evolution, developed different metabolic strategies, i.e., physiological and biochemical adaptations in order to survive, develop, and reproduce under cold stress. The investigations on biochemical and physiological adaptations to low and freezing temperatures of these insect species attracted the attention of many researchers so that in this field over 1500 papers were published by 1986.[91,92] Numerous data on the mechanism(s) through which insect species of polar and temperate regions survive cold stress resulted in a number of review articles from the late 1950s[93] and early 1960s[94] until the late 1980s.[95] Keeping in mind the extensive literature concerning cold hardiness in insects, in this short historical survey only superficial data on the reorganization of the intermediary metabolism during cold will be presented. The intermediary metabolism, during exposure to low, subzero and freezing temperatures, is most extensively studied in one species of Diptera (fam. *Tephritidae*): gallfly *Eurosta solidaginis*, a freezing tolerant species, overwintering as third instar larva, which ranges between central Canada (54°N) to the Gulf Coast Region in the U.S. (27°N). This species was studied extensively by many investigators (about 50 papers), serving as a model in winter hardening studies, commonly referred to as the "Drosophila of cold hardiness."[96] As early as 1950, Salt[97] discovered glycerol in this species and attributed to this substance a cryoprotective role which was confirmed by subsequent investigations on many insect species. However, some 20 years later Morrissey and Baust[98] established, using this species as a model, that glycerol is not the only cryoprotective substance in overwintering larvae; they have a multifactorial cryoprotective system, glycerol-sorbitol-trehalose. Subsequent detailed studies on changes in intermediary metabolism in *E. solidaginis* under the action of low and subzero temperatures showed that the fat body glycogen is the main source of all three components of the multifactorial cryoprotective system. The maximum glycerol synthesis was found between 5° and 0°C and that of sorbitol at about −10°C. These changes are associated with changes in key enzymes of carbohydrate metabolism. Proline, an amino acid which was ascribed a cryoprotective role, also increased, reaching its maximum at −5°C. In the course of gradual acclimation of these larvae from 15° up to −30°C, aerobic metabolism becomes anaerobic at about −10°C. It is surprising that during the acclimation mentioned proteins and total glyceride reserves do not change.[99]

In addition to the study of metabolic changes during the acclimation process in larvae from natural populations from August till April, levels of cryoprotectants (glycerol, sorbitol, sugars), fuel reserves (glycogen, glycerides, proteins), and lactate were monitored and correlated with the ambient temperature profile.[100]

The investigations concerned with primary cryoprotectants at the level of population of this species are of particular interest and were started in the 1970s by Baust and his research fellows.[101] These authors suggest that northern and southern populations of *E. solidaginis* utilize quantitatively different adaptive strategies during winter.

In addition to changes in intermediary metabolism, a direct correlation between low temperature acclimation and increase in water "bound" by soluble subcellular components was proposed in this species,[102] which presents physiological adaptation to low temperatures.

As early as the 1970s in the hemolymph of this species, ice nucleating agents were discovered. They affect the freezing of extracellular fluids and enable the survival of this freezing tolerant species during the winter.[103] However, these extensive studies in the species *E. solidaginis* undoubtedly contributed to the discovery of some links of cold hardiness mechanism in insects.

The data obtained on many other insect species besides *E. solidaginis* indicate the existence of different strategies, developed in the course of evolution, in order to enable insects to survive cold stressory temperatures.

1. Carbohydrates, Lipids

Trehalose is a disaccharide, found in the fat body and hemolymph of the majority of insects investigated thus far. Its discovery in the late 1950s by Wyatt and Kalf[104] contributed significantly to the study of stress-induced changes in the intermediary metabolism in insects.

In view of the particular role this sugar plays in insect physiology, especially under stressory conditions, we shall give an outline of the studies concerned with stress in which this metabolic parameter was monitored and well documented. Wyatt[105] reported that, as a result of experimentally caused injury, which might be considered stressful, in pupae of *Hyalophora cecropia* and *Samia cynthia*, trehalose concentration in the hemolymph increases sharply. In a parallel study, Steele[30] suggested that this metabolic parameter is controlled by the CC, i.e., by neurohormone(s). The involvement of neurohormone(s) in the elevation of trehalose level under stress has been confirmed by the isolation and chemical identification of a hyperglycemic hormone in *P. americana*.[106] Subsequent studies indicated that different stressors (e.g., handling, anaesthesia, capture, exercise) induce, dependent on the insect species, the elevation of either sugar or lipid level in the hemolymph. Thus in *P. americana*, for example, the excitation caused hypertrehalosemia.[56]

However, stressors to grasshoppers induce hyperlipemia in *Locusta* as the response to handling[107] and in *Schistocerca* as the response to heat and chemical stress.[59] These changes are correlated with changes in octopamine concentration, which is considered as the first link in the response to stressors, playing the role of neurohormone at the start of the exposure and then that of neuromodulator.[57] It has been recently shown that some insect species such as the house cricket *Acheta domesticus* respond to the action of stress by both hypertrehalosemia and hyperlipemia.[108] The authors propose that hyperglycemia is induced by octopamine, whereas hyperlipemia at the start of the exposure is directly induced by octopamine and later is associated with adipokinetic hormone, the release of which is stimulated by this biogenic amine.

Studies on the trehalose concentration during starvation, which can be considered also as a stressful factor, demonstrated that in some insects trehalose concentration is increased, e.g., in the tobacco hornworm *Manduca sexta*,[109] whereas in others, e.g., in the larvae of *Rhycosciara americana*[110] and in the larvae of the beetle *M. funereus* trehalose concentration remains almost unchanged.[111] Subsequent investigations showed that in the larvae of *Manduca sexta* during starvation the CC release a neurohormone which stimulates the activation of glycogen phosphorylase, which is one of the key enzymes in the transformation of the fat body glycogen into trehalose.[112] At present, data are not available to indicate why, under starvation stress in the larvae of *R. americana* and *M. funereus*, this metabolic parameter remains relatively stable. Hayakawa and Chino[113] reported that the activation and inactivation of glycogen phosphorylase as well as its synthesis in the fat body are dependent on environmental temperature to which diapausing larvae of the silkworm *Philosamia cynthia* were exposed. Environmental temperature of 20° to 25°C causes a drop in trehalose concentration, whereas lower temperatures induce the elevation in the concentration of this disaccharide.

In the larvae of *M. funereus* trehalose concentration is dependent on the phase of the annual cycle. The concentration of this sugar is high during the winter, as a consequence of the effect of low temperatures.[7] The authors assume that trehalose plays a significant cryoprotective role, which was observed also in other species.[114] Short-term exposure of *M. funereus* larvae to subzero lethal temperatures ($-10°C$) produces a different response at the level of glycogen and trehalose, dependent on the phase of the annual cycle. If larvae of this beetle live and feed on the artificial diet the larval development is shortened to about 6 months as compared to larvae collected from natural microhabitat (under the bark of oak trees) when the development lasts for 3 to 4 years. Appreciable differences were found between larvae grown on the artificial diet and those from natural habitat both at the level of the protocerebral medial neurosecretory cells and at the level of metabolic parameters studied. However, by comparing trehalose concentration of larvae grown on artificial diet and that of those from natural habitat a slight increase in the former is observed, which confirms the findings that this parameter in this species is not affected by environmental factors.[115] Still, one of many questions regarding the complex phenomenon of stress remains to be solved: why, in some species, is the trehalose concentration in the hemolymph a relatively stable metabolic parameter, whereas in others it is unstable under the effect of stressors?

2. Free Amino Acids

Further historical outline on the reorganization of intermediary metabolism in the course of the process of stress will be concerned with amino acids. Data on changes in amino acids under the effect of a variety of stressors are scarce, and little is known about the physiological significance of these changes. The first data concerning the effect of stressors and changes in proline concentration were reported for the species *P. americana* in the late 1950s.[116] In the early 1960s these data were confirmed. Namely, it was established in the same species that DDT and physical stress cause the decline of proline.[117] The decrease in the concentration of this amino acid was found to be influenced also by other stressors: e.g., by dieldrine in *Schistocerca gregaria*;[118] under the effect of stressful temperatures in winter larvae of *M. funereus*;[119] due to the effect of several insecticides and enforced walking in cockroaches.[120] In the last mentioned paper the authors reported on the increase in the concentration of the amino acid taurine in the nervous tissue and hemolymph (250%). It was proved that taurine inhibits spontaneous activity of the nervous system and hence it was suggested that it might be one of the neuroactive substances, formed in response to physical and chemical stressors, which causes the paralysis and death of insects. Amino acid L-leucine and its metabolite isoamyline were attributed a similar role in poisoned silkworms. However, there is still debate over what substances cause paralysis of insects under the effect of stressors.[121]

Ivanović et al.[119] correlate the decrease in proline concentration, induced by stressful temperatures, with its role as an energy source. This role was extensively studied in tsetse fly (*Glossina*) and Colorado potato beetle (*Leptinotarsa decemlineata*) during the flight and reviewed in detail by Bursell.[122] Ivanović et al.[119] studied qualitative and quantitative composition of free amino acids in the hemolymph of *M. funereus* larvae living under natural conditions and those exposed to constant temperatures (0, 8 and 23°C) and found changes in all 16 amino acids examined. These changes depend on the phase of the annual cycle and temperature of acclimation. The most abundant amino acid in the winter is proline. Its concentration falls in May and tends to rise in September.[24] The authors maintain that proline plays a role not only as an energy source but also plays a cryoprotective role in this species. The increased proline concentration in the winter was found in several insect species: in diapausing pupae of *Antherea pernyi*,[123] in larvae of *Eurosta solidaginis*,[99] in the southwestern corn borer and European corn borer,[124] and in diapausing females of Colorado potato beetle.[125] The researchers mentioned also suggested the cryoprotective role of proline. Lefevere et al.,[125] argue that proline in diapausing females of Colorado potato beetle has a solely cryoprotective role and is not an energy source.

On the basis of comparative studies on the neuroendocrine system and free amino acids in the hemolymph of *M. funereus* larvae, deriving from different phases of the annual cycle and exposed to constant temperatures, Ivanović et al. [24,119] suggested that free amino acids, particularly proline, alanine, histidine, and tyrosine are controlled by the neuroendocrine system. In *L. decemlineata* Weeda[126] proved that proline is controlled by some factor from CC probably by adipokinetic hormone (AKH).

In addition to mentioned stressory factors some other stressors may be involved in substantial changes in amino acid pool. Thus, starvation in *Schistocerca gregaria*,[127] and x-radiation in *Ceratitis capitata*[128] result in the increase of the pool. Hill et al.[127] consider that the increase in amino acid pool is the consequence of protein degradation during starvation. Boctor,[128] however, argues that this pool is increased due to x-irradiation because amino acids are not used for protein synthesis. But there is no doubt that only scientific quest in the future will provide the right answer to the question why in some species the amino acid pool increases under stressory conditions.

3. Proteins

During the process of stress appreciable qualitative and quantitative changes of proteins occur. These protein changes represent major links in the protective mechanisms.

a. Heat Shock Proteins

Of particular general interest are proteins, which are synthesized in all the cells of all living beings studied to date, from bacteria to humans, in response to the upper sublethal temperatures. Conversely, they are synthesized under the effect of heat shock and in literature are commonly referred to as heat shock proteins (hsp). However, these proteins are synthesized as the consequence of a variety of chemical and other factors (e.g., heavy metals, dinitrophenol, ethanol, recovery from anoxia) which was well documented by Ashburner and Bonner.[129] These authors point out that heat shock in most eucaryotic cells, including those of *Drosophila,* induces a small set of genes that code for the appropriately named "heat shock proteins". The synthesis of hsp is increased with the increased temperature so that when their rate of synthesis is the highest, total inhibition of normal protein synthesis occurs. Under the effect of less severe heat shock both processes are underway. When initial temperature is restored, heat shock proteins are eliminated and normal protein synthesis is restored in the majority of cases. In the last few years several reviews, outlining the data on these proteins, were published.[130,131]

Heat shock proteins were first discovered in the cells of the representatives from the class of insects, i.e., in the fruitfly *Drosophila melanogaster,* not long ago.[132] Some ten years before, in another species of *Drosophila, D. busciki,* Ritossa[133] observed puff formations on polythene chromosomes of salivary glands when flies were transferred from 20 to 37°C. Tissiers et al.,[132] researchers who discovered heat shock proteins in *D. melanogaster,* correlated the puff formations on certain chromosomes with the synthesis of heat shock proteins. It has been established that under the effect of temperatures above 31°C in all the cells of this species the same proteins are synthesized. In the past few years the genes of seven major heat shock proteins have been cloned and five of the heat shock proteins have been completely sequenced.[130] Key proteins in *D. melanogaster* fall into 3 groups: (1) hsp83, being synthesized at 25°C in a variety of cell lines; (2) hsp70, most frequently referred to as the major one on the grounds that it is synthesized under the most extreme conditions of heat shock; and (3) small hsp, being synthesized not only in response to high temperatures but also during some phases of normal development and governed by ecdysons. This observation supports the hypothesis of Chernysh on the role of ecdysons in the protective mechanisms (see Chapter 4).

It has been observed that, in many plant and animal cells and tissues, homologous proteins are synthesized in response to heat shock. This observation aroused the interest of many researchers to extend studies on these proteins. It should be noted, however, that among organisms studied so far *D. melanogaster* was examined most extensively for exploring the phenomenon of heat shock proteins from the aspect of their synthesis, isolation, and physiological role in the process of heat stress.[130] We shall enumerate hereunder some most important findings on *D. melanogaster* concerning the synthesis and physiological role of these proteins, which were proven to play a crucial role in the protective mechanism of this insect, and probably of other organisms, against heat shock and other stressors: (1) hsp synthesis occurs in the nucleus and is controlled by a separate mRNA which is intensively synthesized under the influence of heat shock.[134] (2) It was established that, during and after recovery from heat shock in *D. melanogaster*, heat shock proteins pass into the cytoplasm. Voluminous data suggest that these proteins affect the establishment of normal transcription and also regulate the synthesis and decay of their own messages by some feedback mechanisms.[135]

Recently, several authors argued that there are qualitative differences in the synthesis of heat shock proteins dependent on the phase of growth and development. More recently, Carvalho and Rebello[136] reported that protein synthesis in response to heat shock differs significantly as a function of cell growth in cell cultures of the mosquito *Aedes albopictus*. Data of Fleming et al.[137] clearly show that there are marked differences in the synthesis of heat shock proteins between young and old fruitflies. Interestingly, in young *Drosophila* 14 heat shock proteins were found compared to about 50 in 45-day-olds. Undoubtedly, the obtained data, specified here above, represent a new approach in the elucidation of heat shock proteins and their role in the stress in insects, and probably also in other living organisms.

b. "Injury" and "Cooling" Proteins

Not only heat shock but also other stressors result in the *de novo* synthesis of proteins which have a protective role. Injury, also considered a stressor, which was observed as early as 1958 by Wyatt[138] in silkworm *Hyalophora cecropia*, enhances RNA synthesis which leads to an increase in the incorporation of labeled amino acids into hemolymph proteins[139] and in other tissues simultaneously.[140] The injury protein was first determined by Laufer.[141] Later, it was found that this protein is synthesized in the hemocytes and was identified as a new apoenzyme of esterase. In addition, it was concluded that injury induced early synthesis of imago proteins by activating the non-active gene.[142] Some 10 years later Marek,[143] having exposed pupae of *Galleria mellonella* to 4°C, observed in the hemolymph of this species so-called "cooling protein", synthesized in hemocytes. Marek's hypothesis holds; cooling proteins (esterase isoenzymes) are synthesized in response to cold temperature as a consequence of premature activation of adult genes, which is also the case with injury proteins.

c. "Antifreeze" Proteins

Subsequent studies on the mechanism of the effect of subfreezing and freezing temperatures on insects revealed the existence of two types of proteins in the hemolymph and other extracellular fluids: (1) thermal hysteresis proteins (THPs), and (2) ice nucleating agents (INA). Phenomenon of thermal hysteresis was discovered as early as the 1960s in the classical studies of cryptonephridial rectal complex of *Tenebrio molitor* larvae. Ramsay[144] found thermal hysteresis in the hemolymph, midgut, and especially in perinephridial space. The proteins responsible for thermal hysteresis were partially purified, but their function was not determined. Not until 1977 did Duman[145] discover thermal hysteresis proteins (THPs) in the hemolymph of freezing-susceptible darkling beetle *Meracantha contracta* (Coleoptera, Tenebrionidae), the species which has no low molecular weight antifreezes. Later, THPs were found and purified in the hemolymph of *Tenebrio molitor*.[146] These proteins, which play the role of antifreezes, were observed chiefly in freeze-susceptible species, particularly in Coleoptera. THPs decrease freezing and supercooling points; they are usually synthesized early in autumn, reaching their maximum in the midwinter, and are eliminated in the spring.[147] Duman and Horwath[147] argue that THPs are adsorbed on ice crystals, thus preventing the freezing of extracellular fluids. However, the discovery of THPs in freeze-tolerant insect species was not expected. Recently, Knight and Duman[148] have shown that THPs are extremely effective in inhibiting ice recrystallization. Their probable function, therefore, in freeze-tolerant species is to counter thermodynamic forces that drive the conversion of smaller to larger (physically damaging) ice crystals over time in frozen state.

Several works are concerned with factors triggering the antifreeze proteins. Temperature is commonly considered the major triggering factor in the synthesis of antifreeze substances.[149] However, Horwath and Duman[150] showed in the species *Dendroides canadensis* that photoperiod and circadian rhythm may also have the dominant role in triggering these hormones.

A group of substances, termed ice nucleating agents (INA), was discovered in the hemolymph and in other extracellular fluids of freeze-tolerant insect species. These substances have a significant physiological role because, by elevating supercooling points, the hemolymph and other extracellular fluids are frozen at higher temperatures, thus avoiding the freezing damage to cells.

Zachariassen and Hammel[151] discovered in the hemolymph INA in three species of tenebrionid beetles. These authors demonstrated that nucleators are susceptible to high temperature, which is indicative of their proteinaceous nature. It is generally accepted that INA are proteins in the majority of cases[95] though it has been shown that in *Tipula trivitata* they are lipoproteins.[152]

After 1976 studies on INA were continued extensively and included other species of freeze-tolerant insects.[153,154] However, it was established, as was the case with THPs, that INA are present also in freeze-susceptible insects but only in summer, whereas in winter they are inhibited. Baust and Zachariassen[154] described for the first time a category of ice nucleators associated with cellular matrix in freeze-susceptible species *Rhaqium inquisitor*. Duman et al.[155] purified ice nucleating protein from freeze-tolerant species *Vespula maculata* and established that amino acid composition is highly hydrophilic. The authors suggested that protein structure may be such as to provide site(s) that order water into embryo crystal(s) reducing the energy barrier to nucleation.

In spite of voluminous investigations, little is known about the site(s) of synthesis, synthesis and regulation of both THP and INA syntheses. Most probably, these issues will be given particular attention in further studies on cold hardiness in insects.

C. CHANGES IN ENZYME ACTIVITIES

Keeping in mind the role of enzymes in the regulation of the course and rate of metabolic processes the investigations on the enzymes under the effect of different stressors are of particular significance.

Enzymes are very flexible to the effect of environmental factors and in response to these factors they either change their concentration or catalytic efficiency or favor evolution of distinct variants of enzymes-isoenzymes.[156]

Out of all stressful factors the impact of temperature was most extensively studied, especially the effect of low temperature. During acclimation to cold many enzymes in different insect species showed positive compensation. Hoffman,[157] summarizing hitherto, obtained results on the effect of cold stress on insect enzymes, points out that enzymes, associated with pathways of energy production, i.e., enzymes involved in the generation and utilization of ATP, increase their activities in cold acclimated insect species, which is a general rule in many ectotherm species.[156] However, there are exceptions to the rule:

ATPase in cold exposed *Tribolium* and *Musca*[158] as well as LDH and SDH in the fat body of cold exposed *Periplaneta americana*[159] are decreased.

We shall note several examples regarding the changes in enzyme activity under the effect of low temperatures and other stressors. Thus, one of the key enzymes in the response of insects to stressors is glycogen phosphorylase from the fat body, which is activated in response to different stressors, e.g., to low temperatures and metabolic stress (starvation, flight). This enzyme was studied in detail, particularly from the standpoint of its regulation. Ziegler and Wyatt[160] were the first to show that diapausing silk moth pupae of *Hyalophora* exposed to 4°C significantly increase glycogen phosphorylase activity, transforming inactive to active form. This resulted in the increased production of glycerol from glycogen at low temperature. On the contrary, Hayakawa and Chino[113] established the dependence between environmental temperatures and enzyme glycogen phosphorylase and glycogen synthase in diapausing silkworm pupae *Phylosamia cynthia*. Phosphorylase kinase and phosphatase may be involved in the activation of glycogen phosphorylase by cold.

The increased activity of glycogen phosphorylase was also observed in the starved larvae of *Manduca sexta*. Siegert and Ziegler[112] found in CC the presence of neurohormone-glycogen phosphorylase activating hormone, which is involved in the activation of this enzyme.

Nonspecific response to many stressors exhibits also esterases. Their synthesis and hence their activity are increased in response to "injury" in *Hyalophora cecropia*,[141] in response to low temperature in *Galleria mellonella*[143] as well as in response to different insecticides in *Musca domestica*.[161]

In insects during the evolution a number of detoxifying enzymes, induced under the effect of many toxic agents, common in natural environment, especially in plants, evolved. Relatively recent application of insecticides prompted the study of these enzymes, which are biochemical bases for the development of resistance in insects. The most extensively studied detoxifying enzymes are cytochrome 450, glutathione-*S*-transferase, and others. In the last few years special attention was paid to the study and purification of the latter in view of the fact that in insects this enzyme is the substitution for the deficiency of cytochrome 450.[162]

Further studies are necessary to determine how environmental factors such as crowding, temperature, insecticide treatment, etc. affect the expression of glutathione-*S*-transferase during the development. This information will be useful in understanding how these enzymes are regulated and at what level insecticide selection might interact with the system.

Studies on digestive enzymes in insects under the effect of stressful factors are scarce. Stressory factors examined are chiefly temperature and, to a much lesser extent, starvation, composition of food, effect of the group, and others. The number of insect species studied is rather small: larvae and adults of phytophagous *Coleoptera*, yellow mealworm *Tenebrio molitor*, cerambycid beetles *Morimus funereus* and *Cerambyx cerdo*, as well as nymphs and adults of cockroach *Periplaneta americana*.

Editors of this book and collaborators have been studying for more than 15 years the effect of stressors, particularly the effect of constant temperatures, on digestive enzymes of phytophagous *Coleoptera*. Our primary objective was to discover the mechanism of the effect of stressors through monitoring the response of the target tissue (the midgut) at the level of digestive enzymes—amylase and protease. There is still debate on the regulation of digestive enzymes. However, it has been shown in *Tenebrio molitor* and *Morimus funereus* that neurohormones from the cerebral complex without corpora allata have a dominant role in the regulation of the midgut amylolytic and proteolytic activities.[163,164] These results, together with the results of comparative studies on the activity of the cerebral neurosecretory system and the activity of the digestive enzymes under different experimental conditions, were the baseline for hypothesizing on the crucial role of neurohormones in the process of stress in the insect species studied.

As early as the 1960s Applebaum et al.[165] discovered in *Tenebrio molitor* larvae a considerable elevation in the level of the midgut proteolytic activity during acclimation of larvae to lower temperature (13°C)—compensatory reaction, whereas the higher temperature (31°C) induced its sharp fall. This temperature was close to the upper limit of survival (34°C) for larvae of this species. Subsequent comparative investigations on the activity of medial neurosecretory cells are indicative of the possible role of neurohormones in the response of *Tenebrio molitor* larvae in the process of stress as well as of the correlation of this response with proteolytic activity.[19]

Our further detailed research concerning the involvement of digestive enzymes (amylase and protease) in the process of stress in larvae of phytophagous insects, primarily in the cerambycid beetle *Morimus funereus*, falls into 2 camps:

1. Monitoring the mechanism of the insect's response at different levels: organism (survival, body weight); neurosecretory system (activity of neurosecretory cells); target - midgut (amylolytic and proteolytic activities) to the effect of temperature and other stressors, as well as to the combined effect of more stressful factors (the majority of relevant references were already quoted). The results of these investigations demonstrate that protease and amylase are very sensitive parameters to the effect of stressors as well as that the response of these enzymes might be taken as the baseline for the evaluation of the intensity of stress.
2. Following up to what extent the outcome of stress is dependent on the species, environmental factors, and physiological state.[7,24,119,166,167]

Das and Das[168] explored the effect of temperature of acclimation on the three digestive enzymes (protease, lipase, amylase) in salivary glands and in gaster of males and females of *P. americana*. Singh and Das[169] investigated this effect on the salivary gland amylase in nymphs and adults. These authors concluded that some enzymes (e.g., lipase) do not show compensatory reaction in animals acclimated to low temperature, whereas other enzymes, dependent on the sex, exhibit different types of compensatory reaction.

The studies on digestive enzymes during the acclimation and acclimatization processes raised the question on the role of compensatory reactions at the level of digestive enzymes in the protective mechanism(s) of presently existing insect species and in the course of their evolution.[165,167-169] Better insight into the mechanism of compensatory reactions will be of applicative significance in the near future for the control of harmful and promoting of useful insect species.

REFERENCES

1. **Selye, H.**, A syndrome produced by diverse nocuous agents, *Nature (London)*, 138, 32, 1936.
2. **Selye, H., Ed.**, *Selye's Guide to Stress Research*, Vol. 1, van Nostrand, New York, 1980.
3. **Sternburg, J. and Kearns, G. W.**, The presence of toxins other than DDT-poisoned roaches, *Science*, 116, 144, 1952.
4. **Wigglesworth, V. B.**, Neurosecretion and the corpus cardiacum of insects, *Pubbl. Staz. Zool. Napoli*, 24, 41, 1954.
5. **Sternburg, J.**, Autointoxication and some stress pnenomena, *Annu. Rev. Entomol.*, 8, 19, 1963.
6. **Ivanović, J., Hladni, M., Stanić, V., and Milanović, M.**, The role of neurohormones in adaptations and stress of phytophagous insects, *Annu. Rev. Sci. Pap. Entomol. Soc. Serb.*, 1, 33, 1979 (in Serb., Engl. summ.).

7. **Ivanović, J., Janković-Hladni, M., Nenadović, M., Frušić, M., and Stanić, V.,** Endocrinological and biochemical aspects of stress and adaptations of phytophagous insects to environmental changes, in *Endocrinological Frontiers in Physiological Insect Ecology,* Vol. 1, Sehnal, F., Zabza, A., and Denlinger, D. L., Eds., Wroclaw Technical University Press, Wroclaw, Poland, 1988, 141.

8. **Chernysh, S. I.,** The response of neuroendocrine system to stressors, in *Hormonal Regulation of Insect Development,* Giljarov, M.S., Tobias, V. I., and Burov, B. N., Eds., "Nauka", Leningrad, 1983, 118 (in Russian).

9. **Cook, B. J. and Holman, M. G.,** Peptides and Kinins, in *Comprehensive Insect Physiology and Pharmacology,* Vol. 11, Kerkut, G. A. and Gilbert, L. I., Eds., Pergamon Press, Oxford, 1985, 532.

10. **Scharrer, B., and Scharrer, E.,** Neurosecretion. VI. A comparison between the intercerebralis-cardiacum-allatum system of the insects and hypothalamo-hypophyseal system of the vertebrates, *Biol. Bull. (Woods Hole, Mass.),* 87, 242, 1944.

11. **Kopec, S.,** Experiments on metamorphosis of insects, *Bull. Int. Acad. Sci. Cracovie B,* 57, 1917.

12. **Scharrer, B.,** Neurosecretion. II. Neurosecretory cells in the central nervous system of cockroaches, *J. Comp. Neurol.,* 74, 93, 1941.

13. **Scharrer, B.,** Insects as models in neuroendocrine research, *Annu. Rev. Entomol.,* 32, 1, 1987.

14. **Hodgson, E. S. and Geldiay, S.,** Experimentally induced release of neurosecretory materials from roach Corpora Cardiaca, *Biol. Bull. (Woods Hole, Mass.),* 117, 275, 1959.

15. **Colhoun, E. H.,** Some physiological and pharmacological effects of chlorinated hydrocarbon and organophosphorous poisoning. *Proc. North Cent. Branch Entomol. Soc. Am.,* 14, 35, 1959.

16. **Davey, K. G.,** The release by enforced activity of the cardiac accelerator from the corpus cardiacum of *Periplaneta americana, J. Insect. Physiol.,* 9, 375, 1963.

17. **Kater, S. B.,** Cardioaccelerator release in *Periplaneta americana* (L.), *Science, N.Y.,* 160, 765, 1968.

18. **Vojtenko, N. I.,** Neuroendocrine activity of *Malacosoma neustria, Portheria dispar and Leucoma salicis* during their toxication with several insecticides, Doctoral thesis, Kijev, U.S.S.R., 1968.

19. **Janković-Hladni, M. and Grozdanović, J.,** The effect of temperature on medial neurosecretory cells of *Tenebrio molitor* larvae, *Gen. Comp. Endocrinol.,* 13, 68, 1969.

20. **Ivanović, J., Janković-Hladni, M., and Milanović, M.,** Effect of constant temperature on survival rate, neurosecretion and endocrine cells, and digestive enzymes in *Morimus funereus* larvae *(Cerambycidae: Coleoptera), Comp. Biochem. Physiol.,* 50A, 125, 1975.

21. **Ivanović, J. P., Janković-Hladni, M. I., and Milanović, M. P.,** Possible role of neurosecretory cells: Type A in response of *Morimus funereus* larvae to the effect of temperature, *J. Therm. Biol.,* 1, 53, 1975.

22. **Glumac, S., Janković-Hladni, M., Ivanović, J., Stanić, V., and Nenadović, V.,** The effect of thermal stress on *Ostrinia nubilalis* HBN *(Lepidoptera, Pyrallidae).* I. The role of the neurosecretory system in the survival of diapausing larvae, *J. Therm. Biol.,* 4, 277, 1979.

23. **Ivanović, J., Janković-Hladni, M., Stanić, V., and Milanović, M.,** The role of cerebral neurosecretory system of *Morimus funereus* larvae *(Insecta),* in thermal stress, *Bull. LXXII Acad. Serbe Sci. Arts, Cl. Sci. Nat. Math., Sci. Nat.,* 20, 91, 1980.

24. **Ivanović, J., Janković-Hladni, M., Stanić, V., Milanović, M., and Nenadović, V.,** Possible role of neurohormones in the process of acclimatization and acclimation in *Morimus funereus* larvae *(Insecta).* I. Changes in the neuroendocrine system and target organs (midgut, haemolymph) during the annual cycle, *Comp. Biochem. Physiol.,* 63A, 95, 1979.

25. **Norman, Ch. T.,** Release of neurohormones in the blowfly *Calliphora vicina* with the respect to insecticidal action, in *Insect Neurobiology and Pesticide Action, Neurotox '79, Soc. Chem. Ind., London,* 305, 1980.

26. **Janković-Hladni, M., Ivanović, J., Nenadović, V., and Stanić, V.,** The selective response of the protocerebral neurosecretory cells of the *Cerambyx cerdo* larvae to the effect of different factors, *Comp. Biochem. Physiol.,* 74A, 131, 1983.

27. **Ivanović, J., Janković-Hladni, M., Stanić, V., and Kalafatić, D.,** Differences in the sensitivity of protocerebral neurosecretory cells arising from the effect of different factors in *Morimus funereus* larvae, *Comp. Biochem. Physiol.,* 80A, 107, 1985.

28. **Keeley, L. L. and Hayes, T. K.,** Speculations on biotechnology applications for insect neuroendocrine research, *Insect Biochem.,* 17, 639, 1987.

29. **Mayer, R. J. and Candy, D. J.,** Control of haemolymph lipid concentration during locust flight: an adipokinetic hormone from the corpora cardiaca, *J. Insect Physiol.,* 15, 611, 1969.

30. **Steele, J. E.,** Occurence of a hyperglycemic factor in the corpus cardiacum of an insect, *Nature (London),* 192, 680, 1961.

31. **Norman, Ch. T.,** Neurosecretory cells in insect brain and production of hypoglycaemic hormone, *Nature (London),* 254, 259, 1975.

32. **Maddrell, S. H. P.,** Excretion in the bloodsucking bug *Rhodnius prolixus* Staal. I. The control of diuresis, *J. Exp. Biol.,* 40, 247, 1963.

33. **Wall, B. J.,** Evidence for antidiuretic control of rectal water absorption in the cockroach *Periplaneta americana* L., *J. Insect Physiol.,* 13, 565, 1967.

34. **Keeley, L. L.,** Development and endocrine regulation of mitochondrial cytochrome biosynthesis in the insect fat body. Delta (14) acid incorporation, *Arch. Biochem. Biophys.*, 187, 87, 1978.

35. **Starratt, A. N. and Brown, B. E.,** Structure of the pentapeptide proctolin: a proposed neurotransmitter in insects, *Life Sci.*, 17, 1253, 1975.

36. **Stone, J. V., Mordue, W., Batley, K. E., and Morris, H. R.,** Structure of locust adipokinetic hormone, a neurohormone that regulates lipid utilization during flight, *Nature (London)*, 263, 207, 1976.

37. **Gade, G.,** The adipokinetic hormone/red pigment concentrating family: structures, interrelationships and functions, *J. Insect Physiol.*, 36, 1, 1990.

38. **Raina, A. K. and Gade, G.,** Insect peptide nomenclature, *Insect Biochem.*, 18, 785, 1988.

39. **Orchard, I.,** Review—Adipokinetic hormones—an update, *J. Insect Physiol.*, 33, 451, 1987.

40. **Maddrell, S. H. P. and Casida, J. E.,** Mechanism of insecticide induced diuresis in Rhodnius, *Nature (London)*, 231, 55, 1971.

41. **Samaranayaka, M.,** Insecticide-induced release of hyperglycemic and adipokinetic hormones of *Schistocerca gregaria*, *Gen. Comp. Endocrinol.*, 24, 424, 1974.

42. **Granett, J. and Leeling, N. C.,** A hyperglycemic agent in the serum of DDT prostrate American cockroaches. *Periplaneta americana*, *Ann. Entomol. Soc. Am.*, 65, 299, 1972.

43. **Maddrell, S. H. P. and Reynolds, S. E.,** Release of hormones in insects after poisoning with insecticides, *Nature (London)*, 236, 404, 1972.

44. **Maddrell, S. H. P.,** The insect neuroendocrine system a target for insecticides, in *Insect Neurobiology and Pesticide Action, Neurotox '79, Soc. Chem. Ind. London*, 1980, 329.

45. **Norman, Ch. T. and Samaranayka-Ramasamy, M.,** Secretory hyperactivity and mitochondrial changes in neurosecretory cells of an insect, *Cell Tissue Res.*, 183, 61, 1977.

45a. **Orchard, I. and Osborne,** The effects of cations upon the action potentials recorded from neurohemal tissue of the stick insect, *J. Comp. Physiol.*, 118, 4, 1977.

46. **Beament, J. W. L.,** A paralysing agent in the blood of cockroaches, *J. Insect Physiol.*, 2, 199, 1958.

47. **Colhoun, E. H.,** Approaches to mechanisms of insecticidal action, *J. Agric. Food Chem.*, 8, 252, 1960.

48. **Sternburg, J., Chang, S. C., and Kearns, C. W.,** The release of a neuroactive agent by the American cockroach, after exposure to DDT or electrical stimulation, *J. Econ. Entomol.*, 52, 1070, 1959.

49. **Cook, B. J.,** An investigation of Factor S, a neuromuscular excitatory substance from insects and crustacea, *Biol. Bull. (Woods Hole, Mass.)*, 133, 526, 1967.

50. **Cook, J. B. and Holt, G. G.,** Neurophysiological changes associated with paralysis arising from body stress in the cockroach, *Periplaneta americana*, *J. Insect Physiol.*, 20, 21, 1974.

51. **Evans, P. D.,** Biogenic amines in the insect nervous system, *Adv. Insect Physiol.*, 15, 317, 1980.

52. **David, J. C. and Lafon-Cazal, M.,** Octopamine distribution in the *Locusta migratoria* nervous system, *Comp. Biochem. Physiol.*, 64C, 161, 1979.

53. **Goosey, M. W. and Candy, D. J.,** The D-octopamine content of the haemolymph of the locust *Schistocerca americana gregaria* and its elevation during flight, *Insect Biochem.*, 10, 393, 1980.

54. **Wilson, M. H. and Rounds, H. D.,** Stress-induced changes in glucose levels in cockroach haemolymph, *Comp. Biochem. Physiol.*, 43A, 941, 1972.

55. **Mathews, J. R. and Downer, R. G. H.,** Hyperglycaemia induced by anaesthesia in the American cockroach *Periplaneta americana*, *Can. J. Zool.*, 51, 395, 1973.

56. **Downer, R. G. H.,** Induction of hypertrehalosemia by excitation in *Periplaneta americana*, *J. Insect Physiol.*, 25, 59, 1979.

57. **Downer, R. G. H., Orr, G. L., Gole, J. W. D., and Orchard, I.,** The role of octopamine and cyclic AMP in regulating hormone release from corpora cardiaca of the American cockroach, *J. Insect Physiol.*, 30, 457, 1984.

58. **Gole, J. W. D. and Downer, R. G. H.,** Elevation of adenosine 3'5-monophosphate by octopamine in fat body of the American cockroach, *Periplaneta americana* L., *Comp. Biochem. Physiol.*, 64C, 223, 1979.

59. **Davenport, A. P. and Evans, P. D.,** Stress-induced changes in the octopamine levels of insect haemolymph, *Insect Biochem.*, 14, 135, 1984.

60. **Orchard, I.,** Octopamine in insects: neurotransmitter, neurohormone, and neuromodulator, *Can. J. Zool.*, 60, 659, 1982.

61. **Fuzeau-Braesch, S.,** Contribution à l'étude in vivo du rôle neurophysiologique d'amines biogènes à l'aide d'un test de motilité chez un insecte subsocial *Locusta migratoria*, *C. R. Soc. Biol.*, 173, 558, 1979.

62. **Morton, D. B. and Evans, P. D.,** Octopamine distribution in solitarious and gregarious forms of the locust *Schistocerca americana gregaria*, *Insect Biochem.*, 13, 177, 1983.

63. **Kozànek, M., Juràni, M., and Somogyiova, E.,** Influence of social stress on monoamine concentration in the central nervous system of the cockroach *Nauphoeta cinerea (Blattodea)*, *Acta Entomol. Bohemoslov.*, 83, 171, 1986.

64. **Lüscher, M. and Engelmann, F.,** Histologische und experimentelle Untersuchungen uber die Auslosung der Metamorphose bei *Leucophaea maderae*, *Rev. Suisse Zool.*, 62, 649, 1960.

65. **Berry, S. J., Krishnakumaran, A., Oberlander, H., and Schneiderman, H. A.,** Effects of hormones and injury on RNA synthesis in saturnid moth, *J. Insect Physiol.*, 13, 1511, 1967.

66. **Piechowska, M. J.,** Effect of ionizing radiation on the endocrine system in insects, *Bull. Acad. Polon. Ser. Sci. Biol.*, 13, 139, 1965.

67. **Pipa, R. L.,** Supernumerary instars produced by chilled wax larvae: endocrine mechanisms, *J. Insect Physiol.*, 22, 1641, 1976.

68. **Cymborowski, B. and Bogus, M. I.,** Juvenilizing effect of cooling on *Galleria mellonella, J. Insect Physiol.*, 22, 669, 1976.

69. **Bogus, M. I. and Cymborowski, B.,** Induction of supernumerary moults in *Galleria mellonella*: evidence for an allatotropic function of the brain, *J. Insect Physiol.*, 30, 557, 1984.

70. **Cymborowski, B.,** Effect of cooling stress on endocrine events in *Galleria mellonella*, in *Endocrinological Frontiers in Physiological Insect Ecology*, Vol. 1, Sehnal, F., Zabza, A., and Denlinger, D. L., Eds., Wroclaw Technical University Press, Wroclaw, Poland, 1988, 203.

71. **Nenadović, V., Janković-Hladni, M., Ivanović, J., Stanić, V., and Marović, R.,** The effect of temperature, oil and juvenile hormone on the activity of digestive enzymes and corpora allata in larvae of *Cerambyx cerdo (Col., Cerambycidae), Acta Entomol. Jugoslav.*, 18, 91, 1982.

72. **Malczewska, M., Gelman, D. B., and Cymborowski, B.,** Effect of azadirachtin on development, juvenile hormone and ecdysteroid titres in chilled *Galleria mellonella* larvae, *J. Insect Physiol.*, 34, 725, 1988.

73. **Horwath, K. L. and Duman, J. G.,** Induction of antifreeze protein production by juvenile hormone in larvae of the beetle *Dendroides canadensis, J. Comp. Physiol.*, 151, 223, 1983.

74. **Baust, J. G., Rojas, R. R., and Hamilton, M. D.,** Life at low temperatures: representative insect adaptations, *Cryo-Letters*, 6, 199, 1985.

75. **Nordin, J. H., Cui, Z., and Yiu, C. M.,** Cold induced glycerol accumulation by *Ostrinia nubilalis* larvae is developmentally regulated, *J. Insect Physiol.*, 30, 563, 1984.

76. **Pullin, A. S. and Bale, J. S.,** Effects of ecdysone, juvenile hormone and haemolymph transfer on cryoprotectant metabolism in diapausing and nondiapausing pupae of *Pieris brassicae, J. Insect Physiol.*, 35, 911, 1989.

77. **Bhaskaran, G. and Jones, G.,** Neuroendocrine regulation of corpus allatum activity in *Manduca sexta*: the endocrine basis for starvation induced supernumerary larvae moult, *J. Insect Physiol.*, 26, 431, 1980.

78. **Subrahmanyam, B. and Ramakrishnan, N.,** The alteration of juvenile hormone titre in *Spodoptera litura* (F.) due to a baculovirus infection, *Experientia*, 36, 471, 1980.

79. **Rauschenbach, I. Yu., Lukashina, N. S., and Korochkin, L. I.,** Role of pupal-esterase in the regulation of the hormonal status in two *D. virilis* stocks differing in response to high temperature, *Dev. Genet.*, 1, 295, 1980.

80. **Rauschenbach, I. Yu., Lukashina, N. S., and Korochkin, L. I.,** Genetic of esterases in *Drosophila*. VII. The genetic control of the activity level of the JH-esterase and heat resistance in *Drosophila virilis* under high temperature, *Biochem. Genet.*, 21, 253, 1983.

81. **Rauschenbach, I. Yu., Lukashina, N. S., Maksimovsky, L. F., and Korochkin, L. I.,** Stress-like reaction of *Drosophila* to adverse environmental factors, *J. Comp. Physiol.*, 157, 519, 1987.

82. **Rauschenbach, I. Yu.,** Genetic control of hormone production and breakdown during metamorphosis under stress, in *Endocrinological Frontiers in Physiological Insect Ecology*, Vol. 1, Sehnal, F., Zabza, A., and Denlinger, D. L., Eds., Wroclaw Technical University Press, Wroclaw, 1988, 169.

83. **Kato, G. and Sanate, S.,** Early termination of summer diapause by mechanical shaking in pupae of *Autherea jamamai (Lepidoptera: Saturniolae), Appl. Entomol. Zool.*, 18, 441, 1983.

84. **Shimizu, I., Adachi, S., and Kato, M.,** Acceleration of the time of moulting and larval ecdysis by electroshocks in the silkworm, *Bombyx mori, J. Sericult. Sci. Jpn.*, 47, 226, 1978.

85. **Cherynsh, S. I.,** Changes in ecdysone secretion in *Calliphora vicina* and *Barathra brassicae* under the influence of damage, *Biol. Sci.*, 5, 51, 1980 (in Russian).

86. **Bodnaryk, R. P.,** An endocrine-based temperature block of development in the pupae of the Bertha Armyworm, *Mamestra configurata Wlk., Invertebrate Reprod. Dev.*, 15, 35, 1989.

87. **Merrill, R. S., Savit, J., and Tobias, J. M.,** Certain biochemical changes in DDT-poisoned cockroach and their prevention by prolonged anaesthesia, *J. Cell. Comp. Physiol.*, 28, 465, 1946.

88. **Heslop, J. P. and Ray, J. W.,** The reaction of the cockroach *Periplaneta americana* L. to bodily stress and DDT, *J. Insect Physiol.*, 3, 395, 1959.

89. **Glumac, S., Koledin, Dj., and Horvatović, A.,** On the causes of insect perishing at higher temperatures (Partial correction of Bachmetjew curve, 1907), *J. Nat. Sci. "Matica Srpska"*, 42, 144, 1972 (in Serbian, Engl. summ.).

90. **Glumac, S., Janković-Hladni, M., Ivanović, J., Stanić, V., and Nenadović, V.,** The effect of thermal stress on *Ostrinia nubilalis* HBN (Lepidoptera, Pyralidae), *J. Therm. Biol.*, 4, 277, 1979.

91. **Baust, J. G., Lee, R. E., and Ring, R. A.,** The physiology and biochemistry of low temperature tolerance in insects and other terrestrial arthropods: a bibliography, *Cryo-Letters*, 3, 191, 1982.

92. **Lee, R. E., Jr., Ring, R. A., and Baust, J. G.,** Low temperature tolerance in insects and other terrestrial arthropods: bibliography II, *Cryo-Letters,* 7, 113, 1986.
93. **Ushatinskaya, R. S.,** *Principles of Cold Resistance in Insects,* Acad. Sci., U.S.S.R. Press (in Russian), 1957, 314.
94. **Salt, R. W.,** Principles of insect cold-hardiness, *Annu. Rev. Entomol.,* 6, 55, 1961.
95. **Storey, K. B. and Storey, J. M.,** Freeze tolerance in animals, *Physiol. Rev.,* 68, 27, 1988.
96. **Baust, J. G.,** Insect cold hardiness: freezing tolerance and avoidance—the Eurosta model, in *Living in the Cold: Physiological and Biochemical Adaptations,* Heller, H. C., Masacchia, X. J., and Wang, L. C. H., Eds., Elsevier/North Holland, Amsterdam, 1986, 125.
97. **Salt, R. W.,** Role of glycerol in the cold-hardening of *Bracon cephe, Can. J. Zool.,* 37, 59, 1959.
98. **Morrissey, R. E. and Baust, J. G.,** The ontogeny of cold tolerance in the gall fly, *Eurosta solidaginis, J. Insect Physiol.,* 18, 267, 1976.
99. **Storey, K. B., Baust, J. G., and Storey, J. M.,** Intermediary metabolism during low temperature acclimation in the overwintering gall fly larva, *Eurosta solidaginis, J. Comp. Physiol.,* 144, 183, 1981.
100. **Storey, J. M. and Storey, K. B.,** Winter survival of the gall fly larva, *Eurosta solidaginis:* profiles of fuel reserves and cryoprotectants in a natural population, *J. Insect Physiol.,* 32, 549, 1986.
101. **Baust, J. G., Grandee, R., Condon, G., and Morrissey, R. E.,** The diversity of overwintering strategies utilized by separate populations of gall insects, *Physiol. Zool.,* 52, 572, 1979.
102. **Storey, K. B., Baust, J. G., and Buescher, Ph.,** Determination of water "bound" by soluble subcellular components during low-temperature acclimation in the gall fly larva, *Eurosta solidaginis, Cryobiology,* 18, 315, 1981.
103. **Somme, L.,** Nucleating agents in the hemolymph of third instar larvae of *Eurosta solidaginis* (Fitch) *(Dipt., Tephritidae), Norw. J. Entomol.,* 25, 187, 1978.
104. **Wyatt, G. R. and Kalf, G. R.,** The chemistry of insect haemolymph. II. Trehalose and other carbohydrates, *J. Gen. Physiol.,* 40, 833, 1957.
105. **Wyatt, G. R.,** Effect of experimental injury on carbohydrate metabolism in silkmoth pupae, *Fed. Proc.,* 20, 81, 1961.
106. **Siegert, K. J., Morgan, P. J., and Mordue, W.,** Isolation of hyperglycaemic peptides from the corpus cardiacum of the american cockroach, *Periplaneta americana, Insect Biochem.,* 16, 365, 1986.
107. **Orchard, I., Loughton, B. G., and Webb, R. A.,** Octopamine and short-term hyperlipaemia in the locust, *Gen. Comp. Endocrinol.,* 45, 175, 1981.
108. **Woodring, J. P., Mc Bride, L. A., and Fields, P.,** The role of octopamine in handling and exercise-induced hyperglycaemia and hyperlipaemia in *Acheta domesticus, J. Insect Physiol.,* 35, 613, 1989.
109. **Dahlman, D. L.,** Starvation of the tobacco hornworm, *Manduca sexta,* I. Changes in hemolymph characteristics of 5th-stage larvae, *Ann. Entomol. Soc. Am.,* 66, 1023, 1973.
110. **Terra, W. R. and Ferreira, C.,** The physiological role of the peritrophic membrane and trehalose: digestive enzymes in the midgut and excreta of starved larvae of *Rhynchosciara, J. Insect Physiol.,* 27, 325, 1981.
111. **Janković-Hladni, M., Ivanović, J., Stanić, V., and Milanović, M.,** Effect of different factors on the metabolism of *Morimus funereus* larvae, *Comp. Biochem. Physiol.,* 77A, 351, 1984.
112. **Siegert, K. and Ziegler, R.,** A hormone from the corpora cardiaca controls fat body glycogen phosphorylase during starvation in tobacco hornworm larvae, *Nature (London),* 301, 526, 1983.
113. **Hayakawa, Y. and Chino, H.,** Temperature-dependent activation or inactivation of glycogen phosphorylase and synthase of fat body of the silkworm *Philosamia cynthia:* the possible mechanism of the temperature-dependent interconversion between glycogen and trehalose, *Insect Biochem.,* 12, 361, 1982.
114. **Asahina, E. and Tanno, A.,** A large amount of trehalose in a frost resistant insect, *Nature,* 204, 122, 1964.
115. **Ivanović, J., Janković-Hladni, M., Stanić, V., Nenadović, V., and Frušić, M.,** The role of neurosecretion and metabolism on development of anoligophagous feeding habit in *Morimus funereus* larvae *(Col., Cerambycidae), Comp. Biochem. Physiol.,* 94A, 167, 1989.
116. **Corrigan, J. J.,** The metabolism of some free amino acids in DDT-poisoned *Periplaneta americana* L. doctoral thesis, University of Illinois, Urbana, 1959.
117. **Corrigan, J. J. and Kearns, G. W.,** Amino acid metabolism in DDT-poisoned American cockroaches, *J. Insect Physiol.,* 9, 1, 1963.
118. **Kulkorai, A. P. and Mehrota, K. V.,** Effects of dieldrin and sumathion on the amino acid nitrogen and proteins in the hemolymph of the desert locust, *Schistocerca gregaria, Pestic. Biochem. Physiol.,* 3, 420, 1973.
119. **Ivanović, J., Janković-Hladni, M., Milanović, M., Stanić, V., and Božidarac, M.,** Possible role of neurohormones in the process of acclimatization and acclimation in *Morimus funereus (Insecta).* II. Changes in the haemolymph free amino acid content and digestive enzymes activity. *Comp. Biochem. Physiol.,* 71B, 693, 1982.

120. **Jabbar, A. and Strang, R. H. C.**, The effects of chemical and physical stress on the concentrations of aminocompounds in the haemolymph and nervous system of locusts and cockroaches, *J. Insect Physiol.*, 31, 359, 1985.

121. **Tashiro, S., Taniguichi, E., Eto, M., and Mackowa, K.**, Izoamyline a possible neuroactive metabolite of L-leucine, *Agric. Biol. Chem.*, 39, 569, 1975.

122. **Bursell, E.**, The role of proline in energy metabolism, in *Energy Metabolism in Insects*, Downer, R. G. H., Ed., Plenum Press, New York, 1981, 135.

123. **Mansingh, A.**, Changes in the free aminoacids of the haemolymph of *Antherea pernyi* during induction and termination of diapause, *J. Insect Physiol.*, 13, 1645, 1967.

124. **Morgan, T. D. and Chipendale, G. M.**, Free aminoacids of the haemolymph of the southwestern corn borer and the european corn borer in relation to their diapause, *J. Insect Physiol.*, 29, 735, 1983.

125. **Lefevere, K., Koopmanschap, A. B., and De Kort, C. A. D.**, Changes in the concentrations of metabolites in haemolymph during and after diapause in female of colorado potato beetle, *Leptinotarsa decemineata*, *J. Insect Physiol.*, 35, 121, 1989.

126. **Weeda, E.**, Hormonal regulation of proline synthesis and glucose release in the fat body of the colorado potato beetle, *Leptinotarsa decemlineata*, *J. Insect Physiol.*, 27, 411, 1981.

127. **Hill, L., Mordue, W., and Highnam, K. C.**, The endocrine system, frontal ganglion in the female desert locust, *J. Insect Physiol.*, 12, 1197, 1966.

128. **Boctor, I. Z.**, Some effects of radiation on free amino acids of adult female Mediterranean fruit fly, *Ceratitis capitata* Wied. *Experientia*, 36, 36, 1980.

129. **Ashburner, M. and Bonner, J.**, The induction of gene activity in *Drosophila* by heat shock, *Cell*, 17, 241, 1979.

130. **Petersen, N. S. and Mitchell, H. K.**, Heat shock proteins, in *Comprehensive Insect Physiology, Biochemistry and Pharmacology*, Vol. 10, Kerkut, G. A. and Gilbert, L. I., Eds., Pergamon Press, Oxford, 1984, 347.

131. **Sheldon, L. and Berger, E.**, Heat shock protein genes in *Drosophila*, in *Endocrinological Frontiers in Physiological Insect Ecology*, Vol. 1, Sehnal, F., Zabza, A., and Denlinger, D. L., Wroclaw Technical University Press, Wroclaw, Poland, 1988, 347.

132. **Tissiers, A., Mitchell, H. K., and Tracy, U. M.**, Protein synthesis in salivary glands of *D. melanogaster* cells, *J. Molec. Biol.*, 84, 389, 1974.

133. **Ritossa, F.**, A new puffing pattern induced by heat shock and DNP in *Drosophila*, *Experientia*, 18, 571, 1962.

134. **Mitchell, H. K. and Lipps, L. S.**, Rapidly labelled proteins of the salivary gland chromosomes of *D. melanogaster*, *Biochem. Genet.*, 15, 575, 1975.

135. **Lindquist, S., Di Dominico, B., Brugaisky, G., Kurtz, S., Petko, L., and Sandos, S.**, Regulation of heat shock response in *Drosophila* and yeast, in *Heat Shock from Bacteria to Man*, Schlesinger, M., Ashburner, M., and Tissiers, A., Cold Spring Harbor Press, New York, 1982, 167.

136. **Carvalho, M. G. C. and Rebello, M. A.**, Induction of heat shock proteins during the growth of *Aedes albopictus* cells, *Insect Biochem.*, 17, 199, 1987.

137. **Fleming, J. E., Walton, J. K., Dubitsky, R., and Bensch, K. G.**, Aging results in an unusual expression of *Drosophila* heat shock proteins, *Proc. Nat. Acad. Sci. U.S.A.*, 85, 4099, 1988.

138. **Wyatt, G. R.**, The metabolic control of insect development, in *The Chemical Basis of Development*, McElroy, W. D. and Glass, B., Eds., John Hopkins Press, Baltimore, 1958, 807.

139. **Telfer, W. H. and Williams, C. M.**, The effect of diapause, development and injury on the incorporation of radioactive glycine into the blood proteins of the *Cecropia* silkworm, *J. Insect Physiol.*, 5, 61, 1960.

140. **Wyatt, G. A.**, Biochemistry of diapause, development, and injury in silkworm pupae, in *Insect Physiology*, Brookes, V. J., Ed., Oregon University Press, Corvallis, 1963, 23.

141. **Laufer, H.**, Blood proteins in insect development, *Ann. N. Y. Acad. Sci.*, 89, 490, 1960.

142. **Berry, S. J., Krishnakumaran, A., and Schneiderman, H. A.**, Control of synthesis of RNA and protein in diapausing and injured *Cecropia* pupae, *Science*, 146, 938, 1964.

143. **Marek, M.**, On the mechanism of some antimetabolic action on the biosynthesis of "the cooling protein" by pupae *Galleria mellonella*, *Comp. Biochem. Physiol.*, 35, 737, 1970.

144. **Ramsay, J. A.**, The rectal complex of the mealworm *Tenebrio molitor* L. (Coleoptera, Tenebrionidae), *Phil. Trans. R. Soc. B*, 248, 279, 1964.

145. **Duman, J. G.**, The role of macromolecular antifreeze in the darkling beetle, *Meracantha contracta*, *J. Comp. Physiol.*, 115, 279, 1977.

146. **Patterson, J. L. and Duman, J. G.**, The role of thermal hysteresis factor in *Tenebrio molitor* larvae, *J. Exp. Biol.*, 74, 37, 1978.

147. **Duman, J. G. and Horwath, K. L.**, The role of hemolymph proteins in the cold tolerance of insects, *Annu. Rev. Physiol.*, 45, 261, 1983.

148. **Knight, C. A. and Duman, J. G.**, Inhibition of recrystallization of ice by insect thermal hysteresis proteins: a possible cryoprotective role, *Cryobiology*, 23, 256, 1986.

149. **Baust, J. G. and Rojas, R. R.,** Review—insect cold hardiness facts and fancy, *J. Insect Physiol.,* 31, 755, 1985.

150. **Horwath, K. L. and Duman, J. G.,** Photoperiodic and thermal regulation of antifreeze protein levels in the *Dendroides canadensis, J. Insect Physiol.,* 29, 907, 1983.

151. **Zachariassen, K. E. and Hammel, H. T.,** Nucleating agents in the hemolymph of insect tolerant to freezing, *Nature,* 262, 285, 1976.

152. **Duman, J. G., Neven, L. G., Beals, J. M., Olson, K. R., and Castellino, F. J.,** Freeze-tolerance adaptations, including haemolymph protein and lipoprotein nucleators, in the cranefly *Tipula trivittata, J. Insect Physiol.,* 31, 1, 1985.

153. **Zachariassen, K. E.,** Nucleating agents in cold hardy insects, *Comp. Biochem. Physiol.,* 73A, 557, 1982.

154. **Baust, J. G. and Zachariassen, K. E.,** Seasonally active cell matrix associated ice nucleators in an insect, *Cryo-Letters,* 4, 65, 1983.

155. **Duman, J. G., Morris, J. P., and Castellino, F. J.,** Purification and composition of an ice nucleating protein from queens of the hornet, *Vespula maculata, J. Comp. Physiol. B,* 154, 79, 1984.

156. **Hochachka, O. W. and Somero, G.,** Temperature adaptation, in *Biochemical Adaptation,* Princeton University Press, New Jersey, 1984, 355.

157. **Hoffmann, K. H.,** Metabolic and enzyme adaptations to temperature, in *Environmental Physiology and Biochemistry of Insects,* Hoffmann, K. H., Ed., Springer-Verlag, Berlin, 1985, 1.

158. **Anderson, R. L. and Mutchmor, J. A.,** Temperature acclimation in *Tribolium* and *Musca* at locomotory, metabolic and enzyme level, *J. Insect Physiol.,* 17, 2205, 1971.

159. **Das, A. K. and Das, A. B.,** Patterns of the enzymatic compensations in the fat body of male and female *Periplaneta americana, Proc. Indian Natl. Sci. Acad. Part B,* 48, 44, 1982.

160. **Ziegler, R. and Wyatt, G. R.,** Phosphorylase and glycerol production activated by cold in diapausing silkmoth pupae, *Nature,* 254, 622, 1975.

161. **Dauterman, W. C. and Hodgson, E.,** Detoxication mechanisms in insects, in *Biochemistry of Insects,* Rockstein, M., Ed., Academic Press, New York, 1978, 341.

162. **Brealey, C. J.,** Pharmacokinetics of insecticides in insects, in *Neurotox '88, Molecular Basis of Drug and Pesticide Action,* Lunt, G. G., Ed., Excerpta Medica, Amsterdam, 1988, 529.

163. **Janković-Hladni, M., Ivanović, J., Stanić, V., and Milanović, M.,** Possible role of hormones in the control of midgut amylolytic activity during adult development of *Tenebrio molitor, J. Insect Physiol.,* 24, 61, 1978.

164. **Ivanović, J., Janković-Hladni, M., Stanić, V., Nenadović, V., and Milanović, M.,** Midgut of Coleopteran larvae, the possible target organ for the action of neurohormones, in *Neurosecretion and Neuroendocrine Activity, Evolution, Structure and Function,* Bargmann, W., Oksche, A., Polenov, A. A., Scharrer, B., Eds., Springer-Verlag, Berlin, 1978, 373.

165. **Applebaum, S. W., Janković, M., Grozdanović, J., and Marinković, D.,** Compensation for temperature in the digestive metabolism of *Tenebrio molitor* larvae, *Physiol. Zool.,* 37, 90, 1964.

166. **Janković-Hladni, M., Ivanović, J., Stanić, V., Milanović, M., and Božidarac, M.,** Possible role of neurohormones in the processes of acclimation and stress in *Tenebrio molitor* and *Morimus funereus.* I. Changes in the activity of digestive enzymes, *Comp. Biochem. Physiol.,* 67A, 477, 1980.

167. **Ivanović, J., Janković-Hladni, M., Spasić, V., and Frušić, M.,** Compensatory reactions at the level of digestive enzymes in reaction to acclimatization in *Morimus funereus* larvae, *Comp. Biochem. Physiol.,* 86A, 217, 1987.

168. **Das, A. K. and Das, A. B.,** Compensations for temperature in the activities of digestive enzymes of *Periplaneta americana, Comp. Biochem. Physiol.,* 71A, 255, 1982.

169. **Singh, S. P. and Das, A. B.,** Thermal acclimatory responses of salivary amylase of *Periplaneta americana, Experientia,* 33, 168, 1977.

Chapter 3

METABOLIC RESPONSE TO STRESSORS

J. P. Ivanović

TABLE OF CONTENTS

I. INTRODUCTION

The external environment of insects represents a dynamic system in which factors show considerable fluctuations in time (seasons, circadian rhythms) and space (geographical latitude), often endangering insect development, reproduction, and their survival in some regions.

The biological plasticity of insects in response to the effect of particular and complex environmental factors is relatively wide ranging, since the optimal combination of the most important factors needed for a successful insect development is seldom established and usually it lasts for a short time. Regardless of this, the insects develop, reproduce, and maintain their population density at the needed level. Moreover, they are capable of a high degree of adaptive radiation for which success of the individual variability of physiological and biochemical characteristics in insect populations plays a dominant role.[1,2] This variability enables the survival under stress and makes the principal basis for the natural selection (transformation of the ecological valence, changes in gene frequency, formation of new adaptive characteristics).

Having in mind the inconstancy of extrinsic environmental factors, the range of their variability, as well as the differences in insect sensitivity during development, reproduction, and aging, the metabolic changes observed in the processes of stress will be arranged according to the principal factor which evokes the stress: temperature, starvation, flight, and chemical agents.

II. THERMAL STRESS

The temperature range in which, under natural conditions, the insect life is possible comprises temperatures ranging from $-70°C$ (some polar species) to $+50°C$.[3] Owing to the fact that the rate of all chemical reactions—therefore most physiological processes—is temperature dependent, the organisms have adopted during their evolution two basic strategies in response to the effect of environmental temperatures. The birds and mammals have evolved mechanisms (thermoregulation) enabling them to develop relatively independently of the ambient temperatures (endotherms). On the other hand, the development of invertebrates and low vertebrates is directly affected by the environmental temperatures (ectotherms). The latter have evolved a certain plasticity both at the levels of functional proteins (enzymes, hormones) and the metabolism.

In recent years it was shown that there are no sharp borderlines between ectotherms and endotherms since some intermediates were described. Although the insects are generally considered to be ectotherms there is considerable evidence that most species are capable of some degree of thermoregulation. Insects achieve thermoregulation either by varying the extent of heat exchange with the environment or by generating metabolic heat. In some butterflies, many moths, bees, wasps, some dragonflies, and beetles, increases of body temperature are the result of heat produced by flight muscles.

A considerable quantity of research into the effect of temperatures on hemi- and holometabolous insect species in different phases of their development, under natural (acclimatization) and experimental conditions (acclimation) and at different levels of biological organization, has been published.

Owing to the limited space in this book, only a small number of papers dealing with the temperature as a stressor will be mentioned. More information about the effect of temperatures on insects can be found in the capital works of Precht et al.,[4] Hochachka and Somero,[5] Prosser,[6] and others. Additional data related to stressful temperatures will be presented in Chapter 2 of this book together with a detailed survey of the history of stress studies in insects.

The temperature acts on insects as a stressor of diverse intensity when it drops below or increases above the insect's optimum temperature range, which varies not only in different species, but also in populations within a species, as well as in different phases of development. The range of stressful temperatures is not continuous since it is interrupted by optimum temperatures. The upper limit of stressful temperature range represents the upper lethal temperature, while the lower one, the lower lethal temperature. Both the upper and lower lethal temperatures differ in diverse insect species.

Since there are differences in the insect's response to the effect of large-scale temperatures, the metabolic changes occurring during cold and/or heat stress will be dealt with in the following temperature ranges: (1) cold stress, metabolic changes provoked by the subzero temperatures ranging from the lower lethal temperature to 0°C; (2) metabolic changes at suboptimal temperatures ranging from the optimum to 0°C; (3) heat stress, metabolic changes elicited under the effect of sublethal temperatures comprising the range of upper lethal to optimal temperatures. The most investigated mechanism in insects is until now that of the effect of subzero temperatures.

A. COLD STRESS
1. Subzero Temperatures

The capacity of insects to withstand subzero temperatures without perishing was described for the first time by Réaumur[7] in the 18th century, 1736. The pioneer works in this field were done in the middle of this century.[8-11] The above-mentioned publications were related to the discovery and study of the mechanisms leading to cold hardiness in insects, defined as "capacity of an organism to survive exposure to low temperature over extended periods, lasting weeks and months".[12] Investigations into cold hardiness in insects have shown that during the evolution insects have adopted different strategies (including the changes in behavior and metabolism) that ensure their survival at subfreezing temperatures. So the Antarctic species of Collembola, *Cryptopygus antarcticus,* in response to cold stress, adopts a quiescent, partially dehydrated state characterized by an ability to rapidly resume normal activities after thawing commences. This is facilitated by a generally hydrofuge body surface, and an ability to rapidly catabolize accumulated polyols. This strategy shows similarities to that of certain desiccation-tolerant estivating insects.[13] Many insect species developing in the Arctic and those overwintering in a moderate climate under outdoor conditions are exposed to subzero temperatures, periods of severe frost, and intermittent freeze-thaw cycles for prolonged periods. The above stressors evoke in insects disturbances at the molecular level, i.e., in conformation, orientation, and mobility of the membrane lipids, disorder in transmembraneous diffusion and transport, as well as changes in the function of metabolic pathways linked to the membrane. Chilling injuries involve changes of different types of interactions between weak links leading to disruption in higher levels of protein structures. It causes changes in individual functions of structural proteins and enzymes. On the other hand, freezing injuries comprise structural damage caused by formation of ice crystals. At the level of an organism these subzero temperatures elicit changes in the general metabolism and the metabolism of regulatory systems.

a. Carbohydrates, Lipids, Proteins

To prevent or diminish the effect of subzero temperatures, insects have evolved a strategy to synthesize several protective substances (cryoprotectants) of carbohydrate or protein nature before the onset of the winter. Cryoprotectants involved in the carbohydrate metabolism are characterized by low molecular weight, e.g., different polyhydroxylic alcohols (polyols), sugars, and the amino acid, proline. Proteins exhibiting a cryoprotective role are the thermal hysteresis proteins (THPs) and the ice nucleating agent (INA).

The most common polyol discovered in insects until now is the glycerol. It was found in insects not susceptible to ice crystal formation in extracellular fluids[14-18] (freeze-tolerant insects [FT]) as well as in those able to inhibit the crystal formation and/or to depress the freezing point (FP) of body fluids (freezing-susceptible insects [FS]). It is supposed that glycerol exhibits different protective roles: it hinders the reduction of the cell volume below the critical level, below which the reduction of cell volume becomes irreversible. Glycerol enables the cells to bear the depressed cell metabolism and, simultaneously, it affects the shape changes of ice crystals turning their sharp edges into the round ones.[19] The second frequent polyol present in insects is the sorbitol.[20,21] The key role of sorbitol as a cryoprotectant for the FT insects is emerging. The sorbitol, not glycerol, can be readily synthesized when ATP is limiting, as in anoxia of frozen state, and only sorbitol is interconvertible with glycogen to allow modulation of cryoprotectant levels with ambient temperature changes.

In addition to polyols, and often simultaneously, the fat body of insects synthesizes some sugars and proline. The sugars and proline act as stabilizers of the cell membrane hindering the phase transition, a consequence of dehydration, i.e., freezing.[22] The concentration of sugars, believed to play a cryoprotectant role in insects, raises at the beginning of winter. The most frequent sugar found in insects is trehalose,[21,23-26] but glucose[11,27] and sucrose were also discovered.[28] It is interesting to recall that the concentration of low molecular weight materials increases during winter to 4 M.[29]

As triggers for the induction of low molecular weight cryoprotectant synthesis might act the hormones and the processes of development, but also several extrinsic factors like temperature, photoperiod, humidity, food, as well as a combination of extrinsic and intrinsic factors.

Glycogen, present in the highest concentration in the fat body, is the basic carbohydrate utilized for the production of the majority of cryoprotectants. Hence, the role of the insect fat body is of greatest significance for the processes of development and stress, both requiring large sources of energy. From the aspect of evolution, the insect fat body is analogous to the mammal liver, an organ system which, in addition to the hypophysis and the suprarenal glands, plays the key role in the processes of stress.[30]

In the central nervous system of insects are two populations of neurosecretory neurons (NSN) which produce neurohormones responsible for regulation of fat body glycogen synthesis. In orders Diptera and Orthoptera, neurohormones regulating glycogen synthesis were synthesized in the medial NSN of the brain[31-33] (MNSN), while in Lepidoptera, this was done in the NSN of the subesophageal ganglion.[34,35] Some authors suppose that the rate of glycogen synthesis is affected by the corpora allata (CA) activity.[36] Recently, it was reported that in the larvae of *Eurosta solidaginis* juvenile hormone (JH) might control the synthesis of certain cryoprotectants such as glycerol and sorbitol.[37] The JH and its analogues stimulate the synthesis of glycerol in *Chilo suppressalis* larvae, while the ecdysone exerts the reverse effect.[38]

It is noteworthy that in some insects the production of cryoprotectants (polyols) is dependent upon the phase of development. So glycerol was produced in fifth instar larvae of *Ostrinia nubilalis* at 5°C.[39] Larvae of *E. solidaginis* synthesize glycerol only in the third instar larvae which overwinter. Dependent upon the fluctuation of the environmental temperatures, some insect species produced polyols and an increased amount of trehalose several times a year.[40,41] During the spring higher temperatures activate the enzyme glycogen synthetase which, once more, converts sorbitol and the surplus of trehalose, at that time functionally needless, into glycogen. Such economical use of sorbitol and trehalose is very advantageous for all insect species developing on a carbohydrate-deficient diet during the spring. It is assumed that glycerol cannot be converted into glycogen, but it seems likely that it may be transformed into lipids.[22] The primary enzyme of glycogen catabolism, the glycogen phosphorylase(s), is activated under effect of low temperatures.[42,43] Recent studies performed on *Manduca sexta* have shown that there exists a strong indication that glycogen

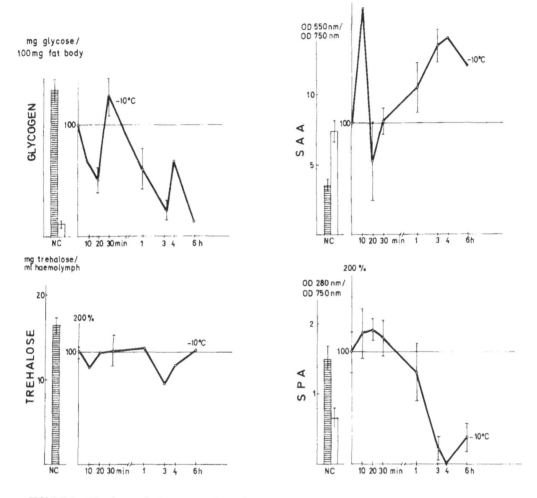

FIGURE 1. The changes in the concentrations of carbohydrates (fat body glycogen, hemolymph trehalose) and digestive enzymes (midgut amylolytic—SAA, midgut proteolytic—SPA activities) in the winter larvae of *M. funereus* exposed to a subzero temperature ($-10°C$).

phosphorylase-activating hormone is stored in the intrinsic cells of CC and as the Golgi zones appear to be active, it is possibly also synthesized[44] there. The release site of the glycogen phosphorylase-activating hormone is unknown, but it could be on the aorta.[45] The results of Gäde and Spring[46] have shown that in the caterpillars of *Romalea microptera* the extract of CC stimulated the fat body glycogen phosphorylase in a dose-dependent manner. On the other hand, extracts of the pars intercerebralis and the subesophageal ganglion had no significant effect on the activation of this enzyme.

The effect of a constant temperature of $-10°C$ and the metabolic changes evoked by this temperature in the glycogen metabolism were followed in *Morimus funereus* larvae used as a model system for studies of the effect of subzero, suboptimum, and high temperatures in the processes of acclimatization and acclimation. Larvae of this cerambycid, collected in winter (termed winter larvae), were frozen at $-10°C$ and then reanimated at $+30°C$. Results presented in Figure 1 show differences in larval response to this stressor in relation to the glycogen and trehalose levels.[47] After 20 min exposure to $-10°C$, glycogen content in the fat body abruptly dropped, whereas after a 6-h exposure it was nearly completely exhausted. The hemolymph trehalose concentration remained at the control level which may be explained by the fact that at this temperature the hemolymph was quickly frozen and its transport stopped.

The larvae of *E. solidaginis* were models in several studies dealing with the effect of subzero temperatures at the molecular as well as the population levels.[48] It was demonstrated that these larvae bear a multiple cryoprotective system. When exposed to subzero temperatures the larvae were able to protect themselves by production of large amounts of sugars (glucose, trehalose), proline, and polyols (glycerol, sorbitol). The breakdown of glycogen leads to an increased concentration of all above-mentioned cryoprotectants.[49] The oxidative processes of the larval metabolism, basically a carbohydrate one, take place in a temperature range of + 15 to − 10°C. In the presence of the key enzymes glycerol kinase and sorbitol dehydrogenase the temperature directly affects the synthetic pathways of the two polyols. The production of glycerol occurs at higher temperatures, i. e., lowering of the temperature to + 5°C induces sorbitol synthesis. The synthesis of sorbitol takes place in oxygen-limited conditions, thus for an anaerobic synthesis of ATP 18% more glycogen is utilized than in an aerobic production of ATP. In larvae of *E. solidaginis* exposed to temperatures ranging from − 10 to − 30°C the oxydative metabolism was inhibited, processes in the fat body inactivated, and the synthesis of cryoprotectants interrupted. It must be stressed that these larvae were able to survive 12 weeks at a temperature of − 16°C. During the first week of larval exposure only 25% disruption in energy metabolism was stated. After 12 weeks spent frozen at − 16°C, the energetic stress of these larvae was very strong in spite of fermentative production of ATP with lactate and alanine accumulation, but with reduced phosphagen reserves in relation to the initial ones (5%). Even this high-intensity stress was not lethal to the larvae since after their reanimation they normally developed, underwent metamorphosis, and 85% of adults emerged.[22]

The insects are able to produce low molecular weight cryoprotectants as well as high molecular weight antifreezes, and therefore protect themselves at subzero temperatures. Proteins which produce a thermal hysteresis (a difference between the FP and melting points [MP]) are well known for their antifreeze function in polar organisms. In fact, proteins with this unique ability, the THPs, were first described by Ramsay[50] in the larvae of *Tenebrio molitor*.

It might be assumed that in the FT insect *Dendroides canadensis* the produced THPs play a protective role against dehydration and probably function as storage proteins.[51] The FS larvae of *Meracantha contracta*, which overwinter protected in decomposed logs, do not accumulate low molecular weight cryoprotectants in the winter, but they proliferate THPs on a seasonal basis.[52,53] The production of THPs only during the winter is synchronous with the depression of the freezing point of the hemolymph. Thus, one may conclude that the THPs are responsible for the supercooling depression which takes place in winter.

The function of THPs in insects is not yet completely clarified, but it is believed that it is more complex. Purified THPs were obtained from *T. molitor* larvae and other insect species, but all have lacked a carbohydrate component.

The THP synthesis is in *M. contracta* and *D. canadensis* evoked by the short photoperiod.[51] Under the effect of low temperatures during November both species have an elevated THP level. It is supposed that the seasonal production of THPs in *D. canadensis* is regulated by the JH.[54] Zachariassen and Hammel[55] demonstrated the biological significance of the extracellular ice nucleators in FT insects. They suggested that INAs probably have the role to establish a protective extracellular freezing at a high subzero temperature. It has been shown that most FT species have INA in the hemolymph. Baust and Zachariassen[56] have shown that in *Rhagium inquisitor* INA were found only in the cell matrix fraction, i.e., they were bound to the cell membrane. The hemolymph INAs from FT insects have been purified and practically characterized. However, the most complete characterization has been done by Neven et al.[57] on a hemolymph lipoprotein INA.

b. Digestive and Other Enzymes

Few data exist on the effect of subzero temperatures on digestive enzymes of insects. In the course of an investigation on the effect of subzero temperatures on larvae of *Morimus funereus* we followed the activity of the two dominant digestive enzymes, the midgut amylase and protease under the effect of − 10°C.[47] Figure 1 shows that there are essential differences in the activities of the two enzymes: after 30 min of larval exposure to − 10°C, the midgut proteolytic activity starts to decrease, and after 4 h of exposure it drops to a negligible value. On the contrary, the midgut amylolytic activity raises after 10 min and after 4 h of exposure for a short time. Afterward, it reaches a higher level than that observed in the controls. This activity level remained unchanged after 6 h of larval exposure too, but at this time 76% of larvae died.

Subzero temperatures activate key enzymes of glycogen catabolism, the glycogen phosphorylase(s).

c. Ionic Concentrations and Free Amino Acids

Evidence was provided that low temperatures may affect the quantitative ionic composition in the insect hemolymph. In the larvae of *E. solidaginis* exposed to low temperatures the concentration of magnesium was markedly elevated, that of sodium and calcium only slightly, while the concentration of potassium was decreased.[22] In contrast to *E. solidaginis*, in *Upis ceramboides* low temperatures evoke a rise in potassium and a drop in calcium, magnesium, and sodium concentrations.[58]

As in other phytophagous Coleoptera in the larval hemolymph of *M. funereus* the most frequent are the cations of potassium and magnesium. During the winter period (January) the concentration of magnesium drastically increases (Figure 2). When these winter larvae were exposed to 23°C the concentration of magnesium abruptly dropped, while that of potassium was not markedly changed.[59] The ionic concentration of *Tipula trivitata* hemolymph remained unchanged during the winter.[60]

In the ectotherms the subzero temperatures and the limited oxygen supply in tissues have caused some changes at the level of the mitochondria, i.e., their number increased, the pH optimum was altered, as was the utilized substrate, too.[61,62]

The hemolymph pH was not significantly different between the cold (3°C) and warm (23°C) acclimated larvae of *E. solidaginis*. When measured at the low temperature the hemolymph pH increased in both experimental groups. This is common for both the blood and the intracellular pH observed in many poikilothermic animals and it is due to the effect of temperature on the PK_a of physiological buffers.[63]

d. Changes at the Level of the Neuroendocrine System

In experiments concerning the metabolic responses of larval *M. funereus* to the effect of − 10°C, the response of protocerebral and subesophageal NSN has been investigated.

It was supposed that the results obtained by measurements of NSN diameters combined with those of other measurable cytological parameters might serve as a semiquantitative interpretation of cell volumetric behavior and therefore be useful in studies of cell responses to the effect of freezing and thawing, enabling us to discover whether the obtained changes are reversible or irreversible.

Figure 3 indicates that the responses of NSN are different, i.e., the are probably the consequence of different resistance of these cells to the given temperature. The diameter of A_2NSN in the medial protocerebral group was drastically at the experimental temperature, and probably irreversibly decreased. Concerning the protocerebral A_1NSN group, the most susceptible one to the effect of all the factors studied (temperature, starvation, azadirachtin), their diameters decreased at the beginning of the exposure to − 10°C, but after 20 min they started gradually to rise and finally they reached the values of the controls. It is interesting

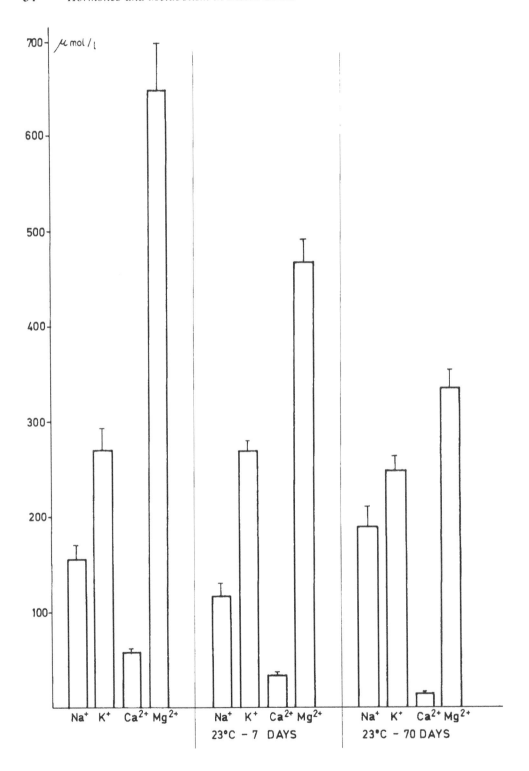

FIGURE 2.   The effect of constant temperature (23°C) on the ionic concentration in the hemolymph of the *M. funereus* winter larvae.

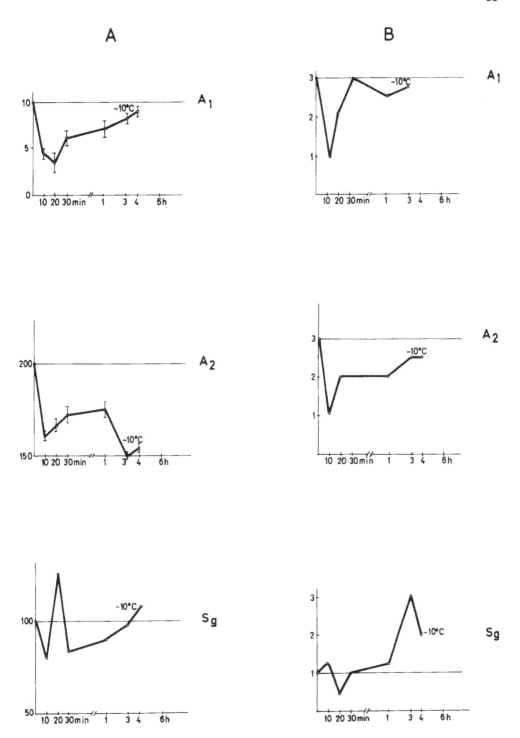

FIGURE 3. The changes in the cell diameter, a × b (A), and in size of nucleoli (B) of protocerebral medial (A₁, A₂) and subesophageal ventromedial neurosecretory neurons (PAF +) in the *M. funereus* winter larvae exposed to a subzero temperature (− 10°C).

to mention that at this temperature, in the medial group of NSN, AZ+ neurosecretory material (AZNSM) was observed as previously stated in the protocerebral lateral group, L$_2$LNSN, under natural conditions during the annual cycle.[64] The diameters of the ventromedial neurosecretory neurons (VNSN) of the subesophageal ganglion increased at the start of the experiment, but afterward their values approached those observed in the control larvae.

2. Suboptimal Temperatures

The range of suboptimal temperatures is very inconstant. It depends on the phase of insect development and the ecology of the species studied. Living under natural environmental conditions the insects are exposed for extended or short periods to suboptimal temperatures. Although very complex and difficult, the studies of the processes of acclimatization are of considerable importance since they represent a prerequisite for an understanding of those metabolic changes taking place not only during the acclimatization processes, but also during the processes of an induced stress, the acclimation to stressful temperatures.

Changes in the insect metabolism evoked by temperatures near 0°C frequently delay insect development and sometimes prevent the metamorphosis causing disturbances in the population dynamics and the ecological equilibrium. The most remarkable changes caused by the effect of suboptimal temperatures were observed in the intermediary metabolism of carbohydrates and free amino acids as well as at the level of the neuroendocrine system. Changes in the protein metabolism with regard to the synthesis of THPs and INA at temperatures near 0°C were discussed in Section II.A.1 dealing with the effect of subzero temperatures.

a. Carbohydrates, Lipids, Proteins

During the processes of acclimatization in the course of the annual cycle the environmental factors, e.g., the temperature, may act on insect populations as a thermal stressor. To adapt to changed environmental conditions insects have evolved mechanisms enabling them to change their general metabolism and particularly the energy metabolism. Figure 4 shows differences in the carbohydrate metabolism (glycogen, trehalose) in *M. funereus* larvae in different seasons and the average monthly temperatures. Seasonal differences observed in the trehalose levels are obvious. It should be particularly emphasized that under natural conditions the hemolymph trehalose concentration starts to rise during the autumn, reaches it highest level in the winter (cryoprotective role), drops abruptly in the spring (period of growth and development), and remains low in the summer. Similar seasonal changes in trehalose levels were reported for other insect species.[65] The concentration of the fat body glycogen also shows seasonal differences, but they are inverse to those observed in trehalose levels. In the larvae from its natural habitat the highest concentration of fat body glycogen was recorded before the beginning of the winter, the lowest one during the winter.[26] Hence, the foregoing experiments show that the winter larvae of *M. funereus* were in a state of natural stress. It is generally accepted that one of the main characteristics of the stress is an activation of the energy metabolism, i.e., raising of the hemolymph trehalose levels, and of glycogenolysis.[66]

The temperature exhibits an influence on the metabolism and the composition of the lipids. In many insect species the development at suboptimal temperatures results in an accumulation of neutral fats and lipids.[4,67] In *Calliphora erythrocephala* and *Sarcophaga bullata* maintained at low temperatures there are changes in the fatty acid composition of the mitochondrial phospholipids, i.e., the amount of long chain unsaturated fatty acids was increased.[68] Downer and Kallapur[69] have shown that the temperature affects the function of the membranes due to the changes arising in the phospholipid composition and cholesterol concentration. According to Singh and Das[70] suboptimal temperatures evoked an intensified protein accumulation in the fat body of *Periplaneta americana*, while in the coxal muscles of the same species changes in protein content were less apparent.

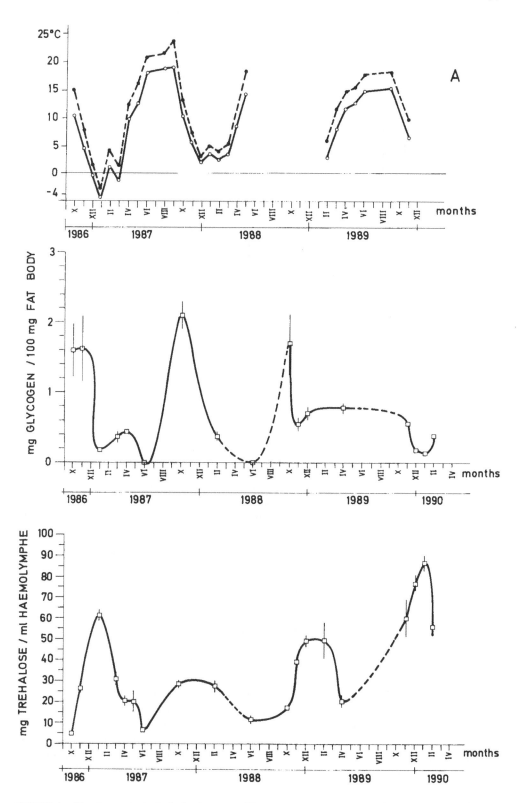

FIGURE 4.   The changes in the fat body glycogen and hemolymph trehalose concentrations in *M. funereus* larvae under natural conditions during annual cycle (acclimatization). A—average monthly outdoor temperatures: ——— at 21 h, ---- 14 h. The data for glycogen and trehalose according to cubic spline method are presented.

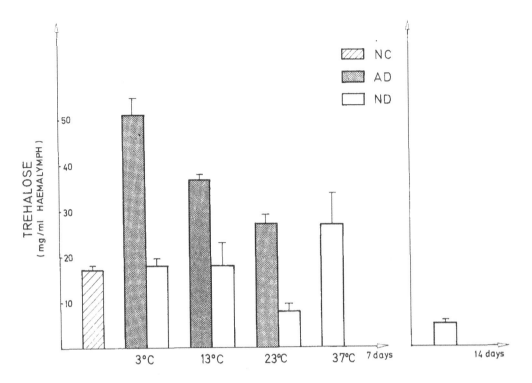

FIGURE 5.   The combined effect of diet (artificial—AD, natural—ND) and different constant temperature on hemolymph trehalose concentration in *M. funereus* larvae (4 d after molting). NC—natural control.

Under natural conditions the effect of environmental factors on the organisms is always complex. To get more insight into the complex action of the environment on the insects, reflected in their responses to it, we performed the following experiment: larvae of *M. funereus* were exposed to thermal stressors (3, 13, and 37°C) and were maintained on an artificial diet of high nutritive value. In comparison to the rate of development observed on the natural diet, the rate of development of those larvae fed an artificial diet was 6.5-fold shorter and the emerged adults were fertile.[71] Figure 5 shows that in larvae fed an artificial diet the temperature gradient was in a negative correlation with the trehalose concentration gradient.[72] Small amounts of glycogen (Figure 6) observed in the fat body of larvae exposed to 3 and 13°C are probably the consequence of a decreased activity of glycogen synthetases, normally inactivated in other insect species during the processes of freezing and thawing,[73] or are the result of an activation of glycogen phosphorylases. In *Hyalophora cecropia* the temperature of 4°C increases the activity of the *a* form of glycogen phosphorylase from fat body more than 50%. This phenomenon has also been demonstrated in *Phylosamia cynthia*.[74] It is reasonable to assume that the activation of glycogen phosphorylases might be regulated by the hypertrehalosemic hormone.[73]

The responses of *M. funereus* larvae fed natural diet to the effect of temperatures differ from those observed in larvae fed an artificial diet. The results obtained indicate that the response of an organism to the effect of a given stressor, in this case the temperature, is dependent upon the effect of other vital environmental factors, such as the quality and the quantity of the food. It was stated that the food directly affects the metabolism of an organism by modifying the mode of the hormonal response.[66]

b. Digestive and Others Enzymes
The digestive system of insects has a specific status since it represents an intermediary between the external and the internal environment. The external temperatures influence this

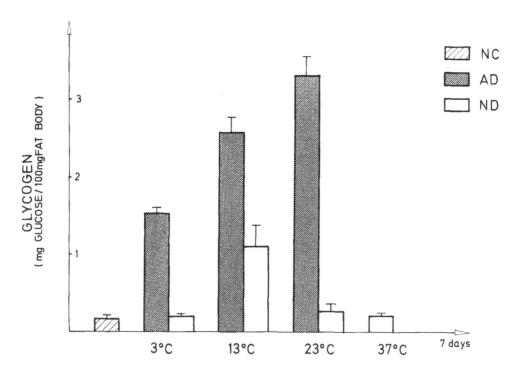

FIGURE 6. The combined effect of diet (artificial—AD, natural diet—ND) and different constant temperature on fat body glycogen concentration in *M. funereus* larvae (4 d after molting). NC—natural control.

system in two ways: they act directly on the rate of digestion, absorption, excretion, and synthesis of the gut hormones, and indirectly via the neurohormones, which in some phytophagous insects regulate the activity of digestive enzymes, the midgut protease, and amylase.[75] Regardless of the fact that the digestive system, i.e., digestive enzymes, are very convenient for studies into the effect of different stressors—in particular, the temperature—only sporadic studies on the above subject have been published. Between the first investigations performed in this field was that of Applebaum and co-workers.[76] The authors have shown that in *Tenebrio molitor* larvae exposed to 13°C the midgut proteolytic activity was elevated by 250% when compared to the controls (larvae at 23°C). Hence, they suggested an existence of a compensatory reaction at the level of this enzyme. Such compensatory reaction was not stated in the activity of midgut amylase. Decreased amylolytic activity *in vivo* at lower temperatures would result in reduced carbohydrate utilization leading to more frequent food intake (behavioral compensation). It was shown that in males and females of *T. molitor* the midgut amylolytic activity was lower at 23 than at 28°C.[77] Following the activities of midgut amylase, protease, and lipase in *Periplaneta americana* reared at 16 and 32°C, Das and Das[78] reported that only in males a partial compensation in proteolytic activity could be stated. The temperature as a stressor did not provoke a compensatory reaction in the activity of lipase neither in the females nor in the males. The amylolytic activity showed high selective compensation under the effect of high temperatures in adults of both sexes.

The suboptimal temperatures affect the midgut proteolytic and amylolytic activities in larvae of *M. funereus* under conditions of an environmental as well as an induced stress (acclimation to suboptimal temperatures).[79] The kinetics of both enzymes shows pronounced fluctuation in their activities at the start of the larval exposure to suboptimal temperatures. Circadian changes in the activity of midgut protease and the concentration of fat body glycogen were also detected in these larvae.[80]

According to the results from literature concerning the rhythmical hormone secretion for the PTTH,[81-83] the control of the circadian rhythm of hemolymph trehalose levels and total sugar,[84,85] and the circadian changes in enzyme concentrations,[86] it appears that the above-mentioned fluctuations in the activity of digestive enzymes in *M. funereus* larvae are also connected with the circadian activity of MNSN that regulate their activities. Afterward, in the course of the experiment, the activity of both digestive enzymes was at a higher or lower level than in the controls, depending on the phase of insect development, season, and the degree of the experimental temperature. There are seasonal differences in the proteolytic activities of *M. funereus* larvae exposed to the same temperature, e.g., to 0°C. In April the proteolytic activity was above, and in September it was below the control level. In this context one question that needs to be addressed is whether these changes were brought about by the changes in the enzyme concentration (quantitative changes), changes in the catalytic efficiency, or changes in the composition of isoenzymes (qualitative changes). The chromatographic separation of larval midgut amylase on DEAE Sephadex A50 has shown two peaks in relation to its activity. However, further electrophoretic procedures used were not adequate for the detection of isozymes[87] (Figure 7).

In the experiment mentioned earlier, simulating more complex environmental conditions (combination of different temperatures and nutritionally different food substrates), the activities of midgut protease and amylase in larval *M. funereus* were followed. It is evident (Figure 8) that in larvae fed an artificial diet the highest level of the proteolytic activity was reached at 13°C, the lowest one at 37°C. After prolonged exposure of the larvae to 23°C the proteolytic activity was significantly elevated. In larvae maintained on a natural diet there were no differences in the proteolytic activities at 3 and 13°C as they were observed at 23 and 37°C. The midgut amylolytic activity of larvae reared on an artificial diet was very low at all the temperatures tested, except at 23°C where it was somewhat higher. Larvae developing on a natural diet showed significantly increased amylolytic activity at 13°C, while at 37°C it dropped to a negligible value. When these larvae were exposed to 23°C for 14 d, the amylolytic activity rose (Figure 9). The increase in the activity of both digestive enzymes observed after prolonged exposure of the larvae to 23°C may well be explained as an attempt of the organism to reestablish a short-lasting metabolic equilibrium (low rate of mortality) (Figure 10). An exposure of the larvae to 23°C lasting longer than 2 weeks resulted in a decreased in glycogen and trehalose contents and an increase in the larval mortality (60%). This high mortality of larvae indicates that their adaptive capacity was exhausted.

The suboptimal temperatures incite changes in the activity of many ''enzymes of metabolic compensation'', used by the ectotherms in an attempt to restore the status quo in the variable extrinsic environment. The changes in the enzyme activities, particularly during the processes of acclimation, are connected with the changes in the concentration of the same enzyme.[5]

Many studies deal with the effect of suboptimal temperatures on an increase in the activities of enzymes, in particular those involved in the energy generation, i.e., enzymes involved in the generation and utilization of ATP such as succinate dehydrogenase, malate dehydrogenase, cytochrome oxydase, glutamate-aspartate transaminase,[88] etc. In other words, the enzymes involved in an aerobic metabolism show remarkable compensatory reactions, whereas the glycolytic enzymes, e.g., lactate dehydrogenase (LDH), do not show compensatory reactions under the effect of stressful temperatures. Even when evoked, these compensatory reactions show an inverse pattern.

However, it was found that in some insects certain temperatures may induce the synthesis of isoenzymes such as the ATPase in *P. americana*[89] and the isoenzyme esterase, the so-called ''cooling protein''.[90] Not only the temperatures induce isoenzyme synthesis, they also differentially change their activities. Mills and Cochran[91] have reported that the four isoenzymes of ATPase in the thoracic muscles of *P. americana* responded differently to the effect

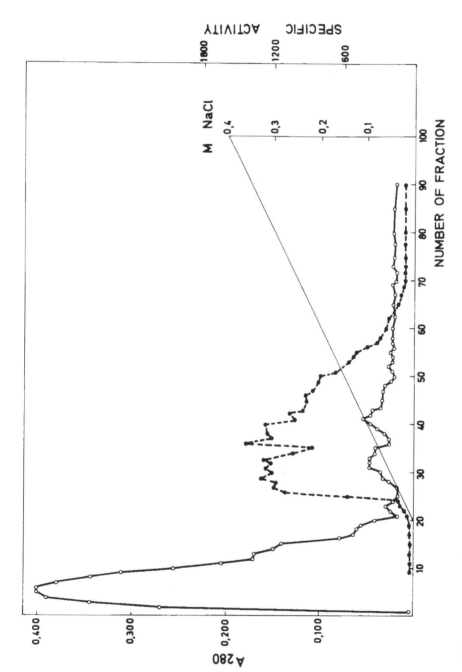

FIGURE 7. Chromatographic dissolving on DEAE— Sephadex A—50 of midgut amylase in *M. funereus* larvae. Absorbance at 280 nm (—○—○—), midgut amylase activity (--○--○--), NaCl concentration gradient (————).

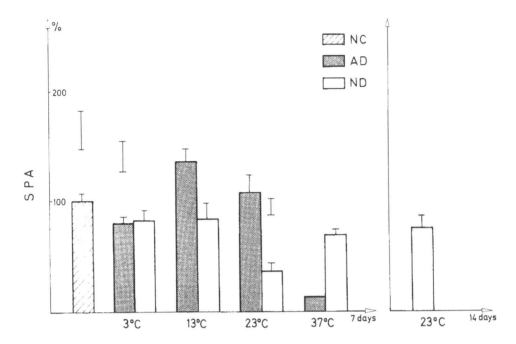

FIGURE 8.   The combined effect of diet (artificial—AD, natural—ND) and different constant temperatures on midgut proteolytic activity in *M. funereus* larvae (4 d after molting). NC—natural control.

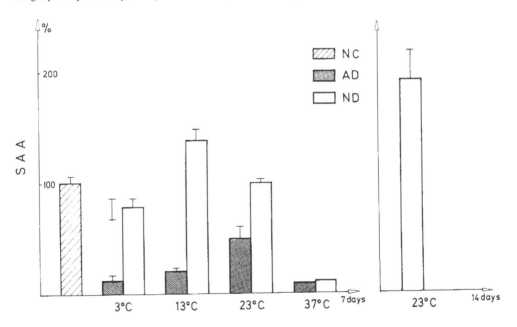

FIGURE 9.   The combined effect of diet (artificial—AD, natural—ND) and different constant temperatures on midgut amylolytic activity in *M. funereus* larvae (4 d after molting).

of 0°C: the ATP-ase IV was inactive, the activity of ATPase II remained unchanged, while the activities of ATPases I and III were somewhat lowered.

It is interesting to recall that in *Musca domestica* the suboptimal temperatures activating the CA[92] were involved in an activation of the insect catatoxic system—microsomal mixed function oxydases—that causes an increase in the general insect resistance to the effect of the environment.

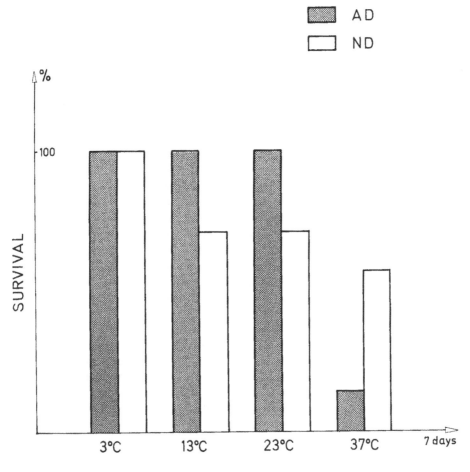

FIGURE 10. The combined effect of diet (artificial—AD, natural—ND) and different constant temperatures on survival in *M. funereus* larvae (4 d after molting).

c. *Ionic Concentrations and Free Amino Acids*

The free amino acid pool of insects and other invertebrates is a center of metabolic activity during a variety of stress responses. The results of our earlier studies concerning the processes of acclimatization of *Morimus funereus* larvae have shown that there are significant qualitative and quantitative differences in the hemolymph-free amino acid content during the annual cycle.[64,93] In insecta, the amino acids are synthesized in the fat body under the control of neurohormones released from CC.[94] During the winter, proline is the dominant amino acid in the hemolymph of *M. funereus* larvae. In the summer its concentration decreases to a negligible value, but in September it starts progressively to rise. In the spring the most frequent hemolymph amino acids are lysine and histidine, whereas in September lysine, histidine, and proline dominate. In the hemolymph of several insect species exposed to low temperatures proline was also present.[12,49,95-97] In larvae of *Eurosta solidaginis* low temperatures provoked an increase in the total titer of free amino acids by approximately 50%. It was predominantly achieved by an elevation of proline (24 mol/g wet wt) and alanine concentrations. The rise in proline concentration is connected with an increased mitochondrial activity, in this species inhibited only at temperatures below −5°C.[49] Some authors assume that proline exerts an effect on the aqueous layer of phospholipids and on the intercalation between the principal groups of phospholipids.[98]

However, not less interesting are the changes in the intermediary metabolism during an induced stress, i.e., changes occurring during the processes of acclimation at suboptimal temperatures. In winter larvae of *M. funereus* maintained at 0 and/or 8°C the concentration of hemolymph glycine, alanine, and tyrosine was significantly elevated, while that of the proline remained at the control level.[64] It was reported that in some insect species during anoxia and ischemia the use of carbohydrates in the metabolic processes resulted in an accumulation of alanine, acetates, pyruvates, and polyols.[22] In the hemolymph of *M. funereus* larvae collected in the spring (May), lysine and histidine are the most abundant amino acids. An exposure of these larvae to 0°C gives qualitative and quantitative changes in the hemolymph-free amino acid content, as a consequence of the drastic rise in alanine content. At a temperature of 8°C glycine concentration was elevated, too, but to a lesser extent.

In the hemolymph of *M. funereus* larvae collected in September the most frequent amino acid is histidine, but at this time the concentration of proline starts to rise. When these larvae were transferred to 0°C, concentrations of alanine and glycine became elevated. The larval transfer to the temperature of 8°C provokes an increase in the concentration of alanine and methionine.[64] Similar results concerning the accumulation of certain hemolymph amino acids (alanine, serine, glycine) were obtained under the effect of suboptimal temperatures.[99] In the larvae of *E. solidaginis* low temperatures have evoked an increase in the concentration of threonine, serine, glycine, and valine. In *Gryllus bimaculatus* under the effect of low temperatures concentration of the total ninhydrin-positive substances was elevated.[67]

d. Changes at the Level of the Neuroendocrine System

Evidence was provided that during the processes of thermal stress in the insects, different metabolic changes in nearly all the organic systems studied took place. The same holds true for the insect regulatory systems under thermal stress, but the provoked changes led to reversible or irreversible disturbances in the hormonal balance of insects.

Investigations into the mechanisms of stress in the vertebrates have shown that stressors elicit an increase in the level of biogenic amines that play an important role in the behavioral responses. Some similarities could be found in the insects. Under the effect of different stressors (temperature, chemical agents, starvation, flight), i.e., during the first phases of the stress, the levels of hemolymph and tissue octopamine were elevated. In *Locusta* an increase in the hemolymph octopamine level starts after 5 min of exposure to the stressor.[100] The hemolymph octopamine regulates the lipid and the carbohydrate metabolism for a short time, thus, in this case it functions as a neurohormone.[101] The release of lipids from the fat body of *Locusta* was caused by the effect of neuropeptides and AKH[102] I. The same could be evoked by a treatment with octopamine. Octopamine activates glycogen phosphorylase in the nerve cord of cockroaches.[103] More recent studies have demonstrated that the biogenic amines are responsible for the glycogen breakdown in the cells of perineurium[66] and therefore for the generation of an additional energy needed for the functions of motor and neurosecretory neurons under stress.

It must be pointed out that at present our appreciation of the activities and functions of different NSN in insects is limited. Hence, it must be stressed that the role of the neuroendocrine system in the metabolic compensatory reactions during the cold and heat stress is as yet unsatisfactorily studied. The first documented case in insects was the report by Clarke[104] on the plausible stimulation of protein synthesis at low temperature (15°C) in *L. migratoria* through neurosecretion, as evidenced by the exhaustion of CC.

Nanda et al.[105] have established that low temperatures and the warming up of *P. americana* change selectively the activity of NSN: A-NSN were activated, and the B- and C-NSN inhibited. In the winter larvae of *M. funereus*, the temperature of 0°C has activated the paraldehyde-fuchsin positive (PF+) VMNSN of the subesophageal ganglion.[79] In the protocerebrum this temperature exhibits different effects on NSN, e.g., the activity of PF+A_1MNSN was decreased and that of PF+A_2MNSN and L_2LNSN was increased. The

L1

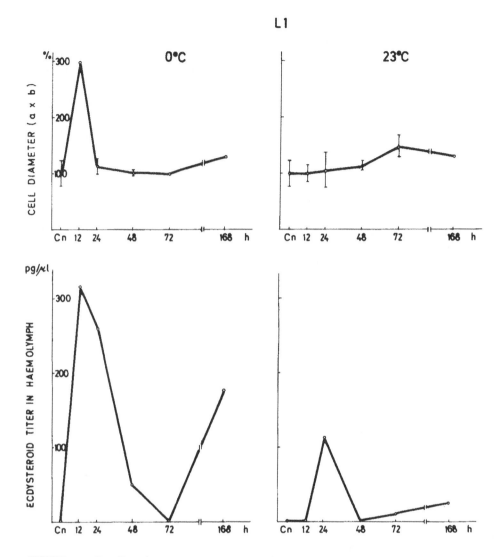

FIGURE 11. The effect of constant temperatures (0, 23°C) on the cell diameter (a × b) of protocerebral lateral (L₁) neurosecretory neurons and ecdysteroid titer in hemolymph of *M. funereus* autumn larvae. (Modified from Ivanović, J. P., Janković-Hladni, M., Nenadović, V., Frušić, M., and Stanic, V., in *Endocrinological Frontiers in Physiological Ecology,* Sehnal, F., Zabza, A., and Delinger, D. L., Eds., Wroclaw Technical Press, Wroclaw, 1988. With permission.)

titer of ecdysteroids in the larval hemolymph and the activity of the digestive enzymes remained at the control levels. On the other hand, in the winter larvae exposed to 23°C the activity of PF + A₁MNSN was raised and the level of ecdysteroids was sixfold higher than in the controls. It must be mentioned that in larvae collected in September and exposed to 23°C the response of NSN was different from that observed in the winter larvae. Here, PF + A₁MNSN were less active, the synthesis of the neurosecretory material (NM) in PF + A₂MNSN was increased, but the transport of NM was slowed down; concerning the lateral group, PF + L₁ and L₂ were activated.[26] In the above larvae there is a correlation between the activity of the lateral group of the protocerebrum (L₁) and the ecdysteroid titer in the hemolymph (Figure 11). It is of particular interest to point out that the concentration of ecdysteroids was significantly increased in larvae exposed to 0°C as well as in those exposed to 23°C. The question was raised why the temperature of 0°C has evoked different responses at the levels of the ecdysteroids in the hemolymph during the autumn and winter. In both seasons the activity of PF + A$_1$MNSN was decreased at 0°C, according to some indirect proofs,[106,107] and using immunochemical methods evidence was provided that in PF + A$_1$MNSN the synthesis of PTTH takes place.[108] In *Manduca sexta* the PTTH is synthesized in the LNSN.[109] Owing to the fact that in the larvae collected in the autumn the activity of L$_1$LNSN and simultaneously the level of ecdysteroids in the hemolymph were increased both at 0 and 23°C, there is a question as to whether or not an intercommunication between L$_1$LNSN and a part of A$_1$MNSN (probably with the same function) was responsible for different responses during different seasons. It might be supposed that different responses during different seasons are connected with seasonal differences in the population of the receptors for PTTH.

In our earlier studies into the mechanisms of acclimatization and acclimation in phytophagous Coleoptera, it was stated that during the annual cycle the histochemical characteristics of the NM in L$_2$LNSN of *Morimus funereus* have changes,[64] i.e., under natural conditions in September L$_2$LNSN produced AZ+ NM, while from November to spring PF+ NM was synthesized. Changes in histochemical characteristics of NM synthesized in L$_2$LNSN and/or MNSN may be evoked under experimental conditions, e.g., under the effect of different constant temperatures. The winter larvae exposed for 10 min to − 10°C produce in MNSN AZ+ NM. However, the autumn larvae (September) that synthesize at this time in L$_2$LNSN AZ+ NM, when exposed to 0°C, produce, instead of AZ+, PF+ NM.

Based on the above results, one may conclude that certain NSN of this and other insect species are capable to produce, under given conditions, either different hormone carriers, different neurohormones, or modifications of the same neurohormone.[93] It seems likely that the above-reported interchange in the production of PF+ and AZ+ NM might be regarded as an indication that the NSN concerned synthesize monoamines, most probably the octopamine. Using the Azan staining technique, Schmid[110] has detected monoamine containing neurones not only in the earthworm, but also in the fly's CNS, and one might also see neurons containing the amines histamine or octopamine that are detected by Falck-Hilarp's histofluorescence or by immunocytochemistry. Fournier and Girardie[111] have found that the A$_1$MNSN, considered by the authors to be bifunctional, synthesize different neuroparsins that might be considered as different neurohormones, not only histochemically different protein carriers of the neurohormones. Evidence was provided that the MNSN of insects, in addition to peptides, produce catecholamines.[112] The A$_1$ and A$_2$ MNSN of *Locusta* contain dopamine. A similar phenomenon was discovered in the neurons of vertebrates indicating that a part of the Dale's principle, "one neuron has the ability to synthesize only one transmitter", needs to be supplemented or changed.

In the winter larvae of *Cerambyx cerdo* exposed to 10°C a decrease in MNSN and an increase in CA activity were observed.[113,114] The activation of the CA under the effect of a suboptimal temperature has great significance, since it was stated that JH increases the biosynthetic capacity of the fat body; thus, it directly affects the energy metabolism in insects.[115] Recent studies of other authors demonstrated that suboptimal temperatures exhibited similar effects on other insect species.[116,117]

B. HEAT STRESS

Terrestrial invertebrates, including the insects exposed to the effects of high temperatures, are simultaneously dehydrated if their integument is not satisfactorily developed. The tolerable duration of exposure to a high ambient temperature has been used as a criterion for heat tolerance. Generally, the upper lethal temperature is changed less by acclimation to heat than is the lower lethal temperature by acclimation to cold. This may be because the optimum temperature is usually very close to the upper lethal temperature. The stress in plants and animals leads to rapid but transient reprogramming of cellular activities to ensure survival during the stress period, to protect essential cell components against heat damage, and to permit a rapid resumption of normal cellular activities during the recovery period. The stress elicited by high temperatures is a complex phenomenon and, consequently, the eventual mortality of different organisms observed under heat stress is evoked by diverse factors. According to Bowler,[118] the primary lesion caused by heat injury is unspecific, whereas the secondary and tertiary lesions are specific. It was shown that in the hemolymph of crayfish transferred to 32°C, the concentration of potassium cations abruptly rises, exhibiting a stressful effect on the CNS and muscles (secondary phenomenon). There were no changes in the hemolymph ionic concentration of the blowfly larvae under the heat stress. High temperatures increase in great numbers the population of reactive molecules in the reactions and they exert influence on the weak chemical bonds. Hence, they affect the protein structure, compartmentalization of proteins (interaction of proteins and membranes), the structure of nucleic acids hormone-receptor binding, etc.[5]

1. Sublethal Temperatures
a. Carbohydrates, Lipids, Proteins

In addition to their role in energy metabolism, the lipids represent structural components of all cellular membranes. Not less important is their role in altering the transfer of ecdysons, JH, pheromones, etc. A relationship between environmental temperature and fatty acid composition in membranes has been suggested for a variety of insect species. The lipid composition of a biological unit will respond to temperature changes so as to maintain optimal lipid function and membrane fluidity. Thus, the most marked effects of temperature changes are likely to be expressed in membrane lipids rather than in total lipid extracts of whole insects. In contrast to the literature concerning the vertebrates, in literature related to the insects there are only few data dealing with changes in the membrane lipids evoked by the temperature. Homeoviscous response of biological membranes,[119] which is supposed to make possible the mechanism of function compensation connected to the membrane such as the passive permeability, ionic transport, the activity of membrane enzymes, and probably the lateral diffusion of proteins, showed that in poikilotherms higher temperatures provoked an increase in the proportion of saturated fatty acids in the major phospholipid classes extracted from membranes.

Downer and Kalapur[69] have shown that in *Schistocerca gregaria* transferred to 45°C, the concentration of cholesterol in the mitochondria of flight muscles was elevated. This reaction affected the function of the membrane and it represented an attempt to neutralize the effect of a high temperature at this level of the biological organization. However, in this species exposed to 45°C there were no differences in phase properties of mitochondrial membranes.

The effect of upper sublethal stressful temperatures which substantially change the metabolism of carbohydrates will be demonstrated on the winter larvae of *M. funereus*. These larvae were exposed to a constant temperature of 30°C for a week. During this experimental period all larvae were alive, but prolonged exposure to 30°C resulted in high larval mortality (80%). The remainders undervent metamorphosis but all the young adults died. Hence, it may be concluded that in the above larvae at the level of the regulatory systems some irreversible injuries were provoked. Figure 12 (A and B) shows that in the winter larvae of *M. funereus* from natural habitat small or medium amounts of fat body glycogen were detected, whereas the concentration of the hemolymph trehalose was relatively high. During the first hours of larval exposure to 30°C the content of fat body glycogen abruptly decreased. After an exposure time ranging from 6 to 12 h some rise in the glycogen content could be stated. Afterward, it sharply decreased and after 72 h the glycogen content in the fat body was almost totally exhausted. After 3 h of larval exposure to 30°C the hemolymph trehalose concentration rapidly decreased, but later on it declined step by step.

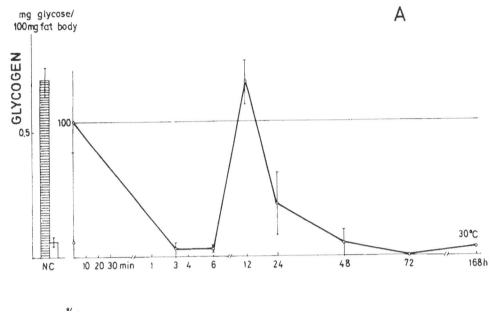

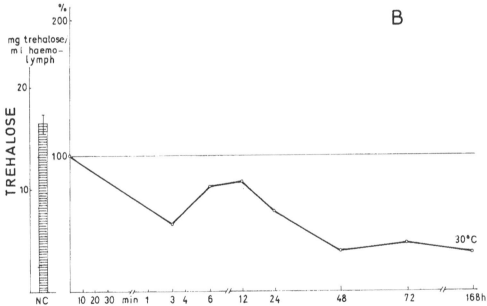

FIGURE 12.   The effect of sublethal temperature (30°C) on glycogen (A) and trehalose (B) concentrations, midgut amylolytic (C) and proteolytic (D) activities in *M. funereus* larvae.

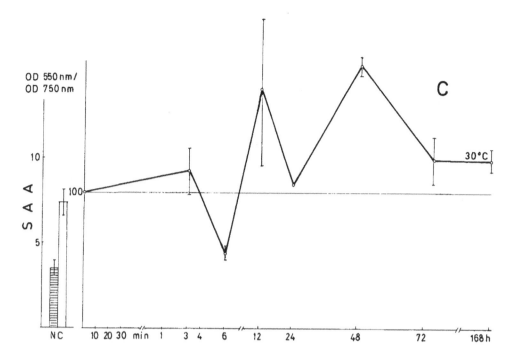

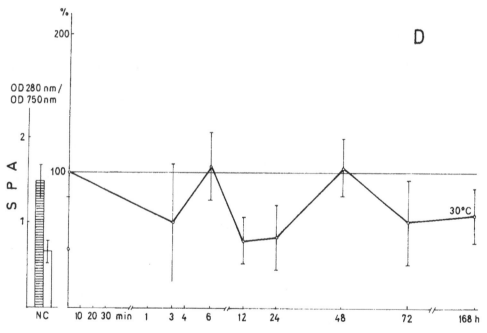

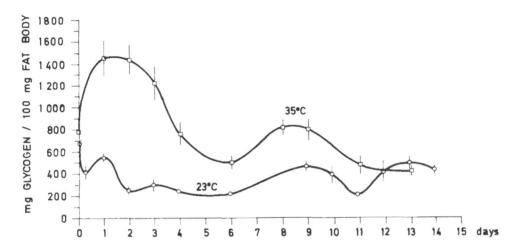

FIGURE 13.   The effect of constant temperatures (23, 35°C) on glycogen concentration during intermolt period in *Morimus funereus* larvae. Data according to cubic spline method are presented.

To get more complex insight into the kinetics of changes in the insect carbohydrate metabolism caused by the effect of stressful temperatures we investigated the changes in fat body glycogen and hemolymph trehalose levels in *M. funereus* larvae (fourth and fifth instar) bred on an artificial diet and exposed to constant temperatures of 23 and/or 35°C during the intermolt period.

Figure 13 shows that the fat body glycogen concentration in the control larvae (23°C) declined after molting, i.e., on day 1 and 2 of the intermolt period. From day 2 to 6, the concentration shows some small fluctuations, but from day 6 to 14 it decreased, reaching its lowest level on day 11. The above results are similar to those observed in the same species during the circadian rhythm of glycogen levels.[80] In comparison with the controls the curve of glycogen concentration in the fat body of larvae exposed to 35°C was markedly different, i.e., it showed two peaks; the first peak occurred between day 1 and 2, the second one between day 8 and 9 during the intermolt period. Immediately after the molt of control larvae the hemolymph trehalose concentration rose. Some oscillations followed in trehalose concentration at a level close to the average values and the obtained plateau in the curve lasted until the end of the intermolt period. The curve of trehalose concentration in larvae exposed to 35°C shows remarkable similarity to the curve of the control larvae, but all the obtained values were considerably lower. It is noteworthy that under the above experimental conditions the trehalose concentration never attained the control level; it remained below it[120] (Figure 14).

Studies of the heat stress in insects have demonstrated that in larvae of *Drosophila hydei* high temperatures provoke impressive changes in the gene activity of the polytene chromosomes of the salivary gland. Subsequent investigations performed on organisms structurally and phylogenetically different have demonstrated that high temperatures evoked in the cells of all the organisms tested vigorous but transient activation of a small number of specific genes previously either silent or active at low levels. New mRNA are actively transcribed from these genes and translated into proteins which are collectively referred to as the heat shock proteins, or hsps. Recently, it was stated that these proteins were synthesized in the cell nuclei not only under the effect of high temperatures, but also under the effect of other stressors, e.g., anoxia, toxic metals, etc. The hsps are functionally homologous but their molecular weights were not identical in different species.[121] Evidence was provided that in different organisms the hsps played a role in thermotolerance, or increased resistance to the stress of elevated temperature. Direct evidence comes from mutant analysis in *Dic-*

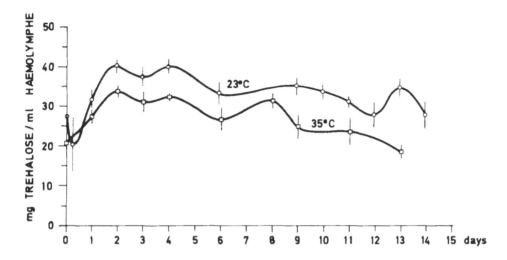

FIGURE 14.   The effect of constant temperatures (23, 35°C) on trehalose concentration during intermolt period in *M. funereus* larvae. Data according to cubic spline method are presented.

tyostelium, which indicates the involvement of small hsps in the development of thermo-tolerance.[122] Recently, pioneering works have appeared in which the dependence between polytene chromosomes and temperature in larvae Chironomus was shown. This dependence is specific for each pond.[123]

b. Digestive and Other Enzymes

In the literature dealing with the cold and heat stress there is relatively little information about the changes in the activity of digestive enzymes provoked by the effect of subzero, sublethal, and suboptimal temperatures.

From Figure 12 (C and D) demonstrating the changes in the activity of digestive enzymes in the winter larvae of *M. funereus* exposed to 30°C, it is evident that at the start of the larval exposure to this temperature the activity of both enzymes shows less or more rhythmic diurnal changes. Afterward, the minor rise in midgut amylolytic activity (low compensatory reaction) reminds one of that observed in the winter larvae exposed to 23°C. After 24 h of larval exposure to 30°C the midgut proteolytic activity declined, but only on one occasion, after 48 h, did it reach the control level. Differential changes in the activity of the digestive enzymes, showing only a low compensatory reaction at the level of midgut amylase, are very similar to those observed in larvae of *M. funereus* collected in February and exposed to 23°C. The observed differences are related to the levels of the activities of both digestive enzymes, i.e., they were higher at 30°C than at 23°C. Owing to the fact that no great differences were seen in the activity of the digestive enzymes at 23 and 30°C, it may be suggested that the sublethal temperatures were less unfavorable for this phytophagous species than were the suboptimal ones. Further investigation in this field must be done to give answers to the question of what reason the sublethal temperatures induce compensatory reactions at the level of midgut amylase in this larvae during the winter and in September, and whether the biological significance of these reactions was to supply larger amounts of carbohydrates necessary in the energy metabolism.

As has already been mentioned, it was shown by Applebaum and co-workers[76] that in larvae of *Tenebrio molitor* a suboptimal temperature of 13°C induces the compensatory reaction at the level of midgut protease. The sublethal temperature of 31°C provoked in the above larvae a sharp decrease in proteolytic activity very quickly, and no activity could be detected after the first week. No compensation for temperature was noted in respect to midgut amylolytic activity at high (31°C) and low (13°C) temperatures.

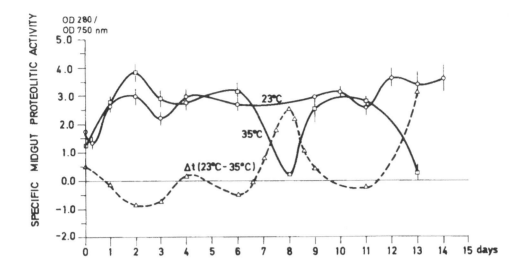

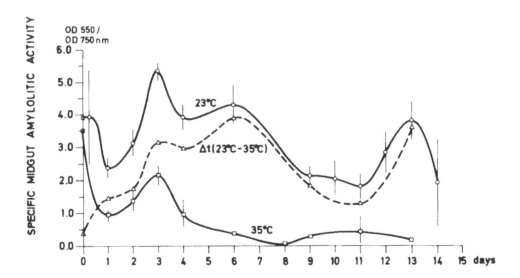

FIGURE 15.   The effect of constant temperatures (23, 35°C) on midgut proteolytic and amylolytic activities during the intermolt period in *M. funereus* larvae. Data according to cubic spline method are presented.

In *M. funereus* larvae the effect of a constant temperature of 35°C on the activity of digestive enzymes during the intermolt period has been studied[120] (Figure 15). Immediately after the larval molt (day 0) the midgut amylolytic activity was relatively high. In larvae maintained at 23°C until the end of the intermolt period (14 d) the midgut amylolytic activity was relatively high and showed fluctuations. However, the amylolytic activity in larvae exposed to 35°C abruptly declined at the beginning of the intermolt period, dropping after day 4 to a negligible value. The midgut proteolytic activity in larvae exposed to 23°C gradually rose at the end of the intermolt period, but in larvae exposed to 35°C it showed compensatory reactions during the first 6 d. Later on, it decreased and on day 13 of the intermolt period it attained a minor level.

Frequently, the suboptimal temperatures activate the CA, i.e., they raise the brain allatotropic activity. The data from literature have shown that under the effect of sublethal temperatures the outcome of heat stress may depend upon the JH metabolism. Bearing in mind that the JH breakdown is caused by the enzyme JH esterase (JHE), it will be interesting to follow the response of this enzyme under the effect of insect heat stress as well as under the effect of certain types of metabolic stress. Such an interpretation is based on the studies of Jones indicating that during the feeding stage neural tissue (primarily the brain) appears directly responsible for regulating JHE activity.[125] Induction of JHE activity by exogenous JH in the postfeeding stage of Lepidoptera suggested that *in vivo* JH induced its own degrading enzyme.[126]

Raushenbach and co-workers[127,128] have demonstrated the important role JH plays in the resistance of the insects to high temperatures. The heat resistance of *Drosophila virilis* pupae is connected with gene expression responsible for the activity of JHE. When this gene is expressed and the JHE activity is high as in larvae of *D. virilis* line 147, the high temperature of 32°C was lethal for this insect. On the contrary, when this gene was not expressed and the JHE activity was low, as in larvae of line 101, high temperature was lethal to the larvae. The hybrids obtained from parents susceptible and resistant to the effect of 32°C were heat resistant. It is therefore evident that the presence of one gene was probably not sufficient to raise the JHE activity. It must be recalled that sublethal temperatures activate the complex of microsomal oxidases, too.

c. Ionic Concentration and Free Amino Acids

High temperatures provoke the changes in osmotic potential of hemolymph and ion concentrations. Under the effect of a sublethal temperature the concentration of Na remained fairly constant and was apparently highly regulated. Concentrations of Ca and K ions were significantly affected by high temperature (40°C) in the hemolymph of larvae *Spodoptera exigua*.[129] The *S. exigua* exposed to high temperature stress showed a dramatic increase in total hemolymph amino acid.

d. Changes at the Level of the Neuroendocrine System

Little information has been available up to now about the effect of sublethal temperatures (particularly those close to upper lethal temperature) on the neuroendocrine system of insects, although the importance of complex and integrated release of chemical messengers (hormones) involved in the protective mechanisms provoked by temperatures and other stressors was evidenced.[130]

The effect of short-term exposure (12 h) to the upper limiting temperature of 45°C on protocerebral A_1MNSN of *Ostrinia nubilalis* larvae has been studied. On the basis of morphometric studies concerning some cytological parameters, oscillatory changes in the activity of A_1MNSN (synthesis and release of NM) were observed. After 1 h exposure to 45°C the activity of A_1MNSN was raised, i.e., the synthesis of NM was elevated, the shape of nuclei changed, and the release of NM increased indicating that the titer of neurohormones, synthesized in the above neurons, might be elevated. A 3-h exposure resulted in a decrease of synthesis and release of NM in the A_1 neurons, and in an increase in the percentage of A_1 neurons with lobed nuclei (preventive inhibition). After 6 h exposure to the same temperature some renewed increase in the activity of these neurons was stated, but the activity was less intensive than at the beginning of the larval exposure to stressor.[131] It is interesting to note that at the start of larval exposure to 45°C the larvae have intensified their locomotory activity, soon replaced by a state of paralysis of the posterior part of the larval abdomen. Cook and Holman[130] believe that the stress paralysis is caused by autotoxic agents of unidentified chemical structure. In this context, an overdosage of biogenic amines, also released in the insects during the first phase of the stress, might be the cause of the paralysis.[132] High temperature and other unfavorable conditions are known to block the secretion of the prothoracicotropic hormone in larvae. *D. virilis*, leading to reduce titers of ecdysone and delayed metamorphosis.[133]

Short exposure to a sublethal temperature evokes frequently reversible responses in the treated insects. So, when post-diapause pupae of *Mamestra configurata* were transferred to 30°C in all the individuals the development was blocked.[134] The author considers that a high temperature causes a reversible stress-like response in pupae of *M. configurata* in which the endocrine system is suppressed and the initiation of development is blocked.

When winter larvae of *Morimus funereus* were exposed to a stressful temperature of 30°C for 7 d no mortality occurred, but after 30 d of exposure the majority of larvae died and the remainders developed into pupae and short-lived adults. The effect of 30°C on the activity of protocerebral MNSN, A_1 and A_2, and VMNSN of the subesophageal ganglion of *M. funereus* larvae is presented in Figure 16 (A and B). One may conclude that during the winter the activity of A_1 and A_2NSN was very high, while that observed in VMNSN was lower. After 6 h exposure to 30°C MNSN activity was decreased. On the basis of morphometric studies it was possible to detect differential responses of different types of the MNSN at the level of some cytological parameters. After 72 h exposure to 30°C the cell diameter of A_2MNSN decreased approximatively by 50%, whereas in a part of A_1MNSN, supposed to be prothoracicotropic, it decreased less than in A_2MNSN. However, at the same time in A_1MNSN the size of nucleoli was the smallest and the accumulation of NM the largest, suggesting a very low activity of these neurons. It must be mentioned that A_1 and A_2MNSN differ in their susceptibility to the effect of this temperature but their responses were similar, only they were at different levels. Regarding the changes in VMNSN and PF+ neurons, they were inverted in comparison to A_1 and A_2MNSN.[26]

III. METABOLIC STRESS

Over the last years in the studies into the mechanisms of stress much attention has been devoted to the changes in the organisms evoked by starvation, deficient nutrition, intensive locomotory activity, and flight. All these processes cause the metabolic stress. They are characterized by enormous energy requirements and when lasting a prolonged time, they lead to the death of the organism.

A. STARVATION AND INCONVENIENT FOOD
1. Carbohydrates, Lipids, Proteins

According to the data from literature concerning the effect of stressful temperatures on the organisms it was stated that for the increased energy requirements, the carbohydrates (fat body glycogen) were mostly utilized. The utilization of fat body glycogen at least in certain insects is under hormonal control, since it is readily depleted by injection of CC extract and converted into trehalose which passes into hemolymph. The CC factor acts directly on the fat body since the glycogenolytic effect and accompanying trehalose efflux can be demonstrated *in vitro*.[135]

Starvation of a 3-d-old last instar larvae of *Manduca sexta* causes a decline in the fat body glycogen concentration. The total amount of sugars in the hemolymph of 48-h starved *M. sexta* was not decreased. Moreover, their level was somewhat elevated. According to these results a decrease in the trehalose concentration might not be a signal for the activation of GP in the fat body, since in the presence of an activated GP trehalose level is high. However, in the starved larvae of *M. sexta* the concentration of hemolymph glucose abruptly decreased. Therefore Ziegler[136] supposed that the glucose concentration might trigger the release of GPAH and consequently the activation of GP. There are direct proofs for the above presumption. When glucose was injected into starved *M. sexta* larvae it hindered the activation of GP, while the trehalose injection did not evoke significant GP activation.[137] In the course of fasting, the larvae must quickly convert their metabolism into catabolism, and intensify their locomotory activity connected with the search for food (additional energy losses). To meet the energy requirements at this time, the GP must be activated and the hemolymph glucose concentration decreased, in this species being an indicator for starvation. Under the effect of prolonged starvation GPs are inactivated and the larvae develop into pupae if they attain the minimum size needed for pupation.

In larvae of *M. sexta* different mechanisms exist which regulate the carbohydrate metabolism during fasting. In the starved adults of *M. sexta* the concentration of fat body glycogen and the hemolymph trehalose decreased, the hemolymph volume was reduced, and the GP was activated. At the same, starvation induces an elevation of the hemolymph lipid levels. It is supposed that an increase in the level of hemolymph lipids provoked by fasting is not regulated by hormones released from CC, but is triggered by the decrease in hemolymph sugar concentration. It seems probable that in starved *Locusta migratoria* the hemolymph lipid level is regulated in a similar way.[138] In the above species starvation causes an intensive gluconeogenesis believed not to be under the control of hormones.[139] In the initial phase of starvation the amount of carbohydrates decreases markedly. A decline in the lipid content starts afterward. In fasting males of *L. migratoria* total carbohydrate concentration strongly decreased. The fall was significant after only 2 d of fasting. The trehalose concentration decreased sharply and after 6 d of fasting it accounted for only 51% of the total carbohydrates.[140] In *Bombyx*, after 2 d of starvation the trehalosemia decreased[141] from 50 to 70%, while in *Anthonomus grandis* the trehalose concentration dropped abruptly.[142]

The above results indicate that during starvation, the primary source of energy utilized is the carbohydrates, normally also indispensable for maintaining the needed carbohydrate level in the hemolymph.

Insect fasting or nutrition with agar evoke a fall in the level of hemolymph trehalose. It was also shown that starvation or feeding an unfavorable, deficient diet may affect the structure of some tissues, e.g., in *Aedes aegypti* fasting provokes drastic reduction of endoplasmatic reticulum in the cytoplasm of enterocytes.[143]

2. Digestive and Other Enzymes

In insects deprivation of food causes several changes at the level of physiological and biochemical parameters, especially those in the functional proteins, the enzymes. It was stated that in fasting larvae of *Morimus funereus* midgut proteolytic and amylolytic activities were significantly decreased.[75] Terra and Ferreira[144] have demonstrated that in starved larvae of *Rhynchosciara americana* the activity of the digestive enzymes (amylase, cellulase, trehalase, aminopeptidase, trypsin) was decreased. The starved larvae of *Nauphoeta cinerea* and *Leucophaea maderae* showed a decline in the midgut proteolytic activity. Under the complex effect of starvation and an unfavorable temperature, in *M. funereus* larvae a significant rise in midgut proteolytic activity was attained (compensatory reaction), while the amylolytic activity decreased.[145]

It was demonstrated that in starved larvae of *Tenebrio molitor* the concentration of the tissue trehalase decreased.[146] Similar results were obtained in the homogenates of starved *Galleria mellonella* larvae.

3. Ionic Concentrations and Free Amino Acids

There was a decrease in hemolymph amino acid concentrations in starved larvae of *Spodoptera exigua*. Ca^{2+} and K^+ ion concentrations were significantly affected by starvation. Concentration of Ca^{2+} ion was decreased, while concentration of K^+ ion was increased.[129]

In fasting *Schistocerca gregaria* the concentration of hemolymph-free amino acids was twofold increased, while that of the proteins was decreased by approximately 50% when compared to the controls.[147] Similar results were observed in *Drosophila* in which, during starvation, after a rapid fall in the concentration of the majority of the hemolymph amino acids their level remained unchanged.[148] In the aphid *Myzus persicae* fed sucrose, the free

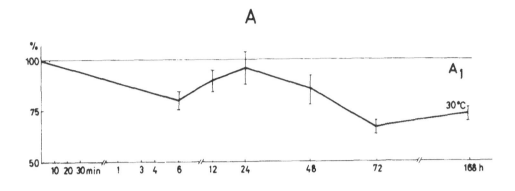

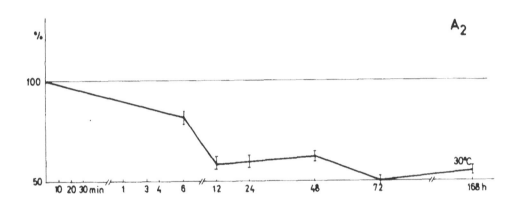

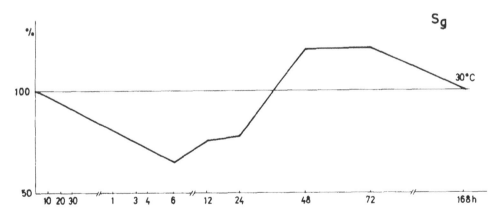

FIGURE 16.   The effect of sublethal temperature (30°C) on cell diameter (a × b) (A), and size of nucleoli (B) on protocerebral medial (A_1, A_2) and subesophageal ventromedial (Sg) neurosecretory neurons (PAF+) in *M. funereus* winter larvae.

B

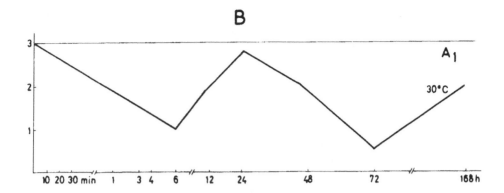

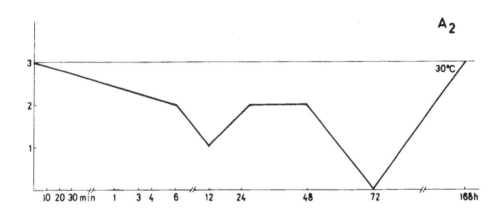

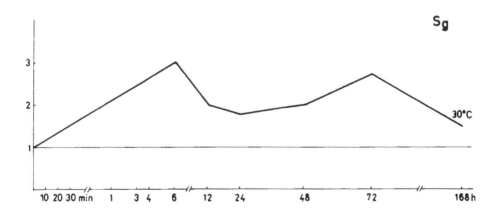

amino acid content of the hemolymph decreased by 50 to 70%. However, the concentration of threonine was at the control level and that of cysteine was somewhat increased.[149] Bosquet[150] reported that in starved *Phylosamia cynthia* the level of lysine, histidine, and glycine was raised, while that of ornithine was at the control level. In short, it may be concluded that during insect starvation only little changes in the concentration of the hemolymph-free amino acids were observed (regulatory mechanisms).

4. Changes at the Level of the Neuroendocrine System

The effect of fasting and deficient nutrition on the insect NSN in the pars intercerebralis of the protocerebrum and on the neuroendocrine system are poorly investigated.

It was shown that deprivation of food slows down the release of the neurosecretory material from A_1NSN of *Bombyx mori* larvae[106] and MNSN of *T. molitor* adults.[151]

The effect of a combined stress, the temperature and starvation observed in the majority of *M. funereus* larvae on both types of MNSN, indicates a drastic decrease in the activity of A_1NSN as well as a slight decrease in the activity of A_2NSN. In addition, some larvae exhibit a compensatory response in both types of NSN, which is particularly marked in A_1NSN. Compensatory response of MNSN under the effect of a double stress might be of particular biological significance for the survival of individuals, i.e., of populations under unfavorable conditions in nature.[152]

Grossman and Davey[153] have shown that in the males of *Glossina austeni* during starvation the number of protocerebral NSN rose from 6 in the fed flies to 20 in the starved ones. The authors also showed that the CC from fed flies is always less than that of the CC from fasting flies. The explanation of the foregoing results was either was a greater release of the neurohormones from the CC in fed flies, or the rate of their release remains approximately constant and the rate of transport into CC from the NSN decreases.

Studies concerning the development of the gypsy moth, *Lymantria dispar* caterpillars, on two natural diets, i.e., on oak and/or locust tree leaves, have demonstrated differences not only at the level of the organism, but also at the level of the neuroendocrine system.[154] The obtained differences were in connection with the differences at the level of population. Figure 17 demonstrates that caterpillars originating from an oak forest and fed locust tree leaves developed longer than normally when fed oak leaves. The prolonged development was selective with regard to the larval instars, i.e., the most prolonged development was observed in the first and the last larval instars. It is also evident that caterpillars originating from a locust tree forest and fed oak leaves developed significantly shorter, especially the first instar larvae. In the fourth instar caterpillars originating from the oak forest and reared on locust tree leaves, the number of the protocerebral NSN, i.e., A_1MNSN and LNSN, was increased in comparison to the controls (larvae fed oak leaves) (Figure 18). The locust tree leaves are an unfavorable food substrate for insects since they probably contain some materials which inhibit insect development. The possible existence of the gypsy moth in nature on the locust tree, whose leaves were consumed by a small number of insect species, is probably due to the physiological and biochemical plasticity of its NSN and also to the fact that similar to all the phytophagous insect species, the gypsy moth has an active system of microsomal oxidases.

In *Manduca sexta* prolonged starvation causes profound changes in the endocrine system, too. After decrease of the JH level after ecdysis, JH titer increases to a level about tenfold higher than the 0.2- to 0.3-ng JH^I equivalents per milliliter seen in 2- to 4-, and 5-g larvae given free access to food. Some authors[155] have suggested that starvation provokes allatotropin secretion and prevents the allatohibin secretion in starved *M. sexta* larvae. A large increase in esterase activity was observed on the second and third days of fasting. In *Galleria mellonella*, too, starvation for 4 days prevents an increase of the JHE activity.[156]

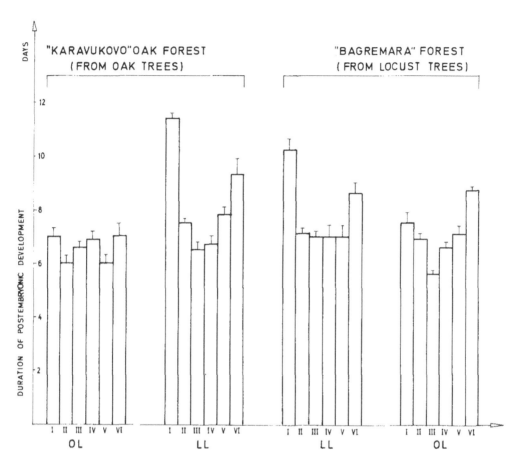

FIGURE 17. Duration of postembryonic development (first to sixth instars) of gypsy moth, *L. dispar*, caterpillars originated from two populations: oak forest ("Karavukovo") and mixed forest (locust tree, oak tree) ("Bagremara"); OL— oak leaves, LL—locust leaves.

B. FLIGHT ACTIVITY

In the course of the insect flight activity the amount of the reserve materials stored in the flight muscles of adults is unsatisfactory to meet the requirements for an enormous energy need, characteristic for the metabolism of flying. Hence, in insects flying a long distance, additional reserve materials accumulated in the fat body (glycogen, triacylglycerol, proteins) were utilized as fuel. The latter were transported from the fat body via the hemolymph into the cells, i.e., the cell organelles.

In the majority of insects species, during an initial phase of the flight activity the carbohydrates were utilized as an energy source. This was evidenced by changes taking place in the concentrations of fat body glycogen and hemolymph trehalose. So, in *Locusta migratoria*, after the first 30 min of flying the trehalose level dropped by 50% but the prolonged flight activity did not affect the trehalose level in the hemolymph. An increase in the trehalose level was followed by a decrease in the fat body glycogen content, which after 2 h of flying decreased to 75%, and by an activation of the GP. After 5 min of flying the percentage of the active enzyme (a) has risen[130,157] up to 25%. Octopamine increases the rate of the GP activity and stimulates the process of glycogenolysis in the nerve cord of the cockroach.[103] At the same time, the rate of glucose oxidation in the flight muscles of the cockroach increases.[158] When octopamine acts on the isolated fat body of the cockroach the glycogenolysis was elevated, which was presumably connected with an increase in the

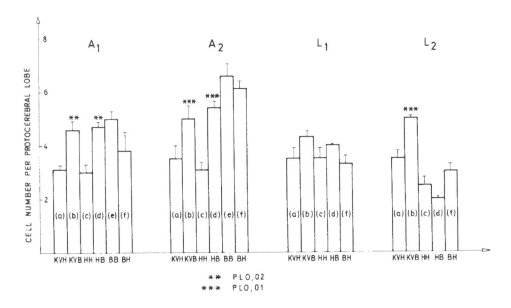

FIGURE 18.   The effect of population and food on the number of protocerebral neurosecretory neurons (medial—A_1, A_2; lateral—L_1, L_2) in gypsy moth, *L. dispar*, caterpillars. KVH—population from oak forest fed oak leaves; KVB—the same population fed locust leaves; HH—population from mixed forest (oak trees) fed oak leaves; HB—the same population fed locust leaves; BB—population from mixed forest (locust trees) fed locust leaves; and BH—the same population fed oak leaves.

hemolymph trehalose level.[159] In the grasshoppers the level of D-octopamine ranged from 33 n*M* during the rest to 173 n*M* observed after 10 min of flying.[160] Evidence was provided that the AKH released during the flight activity stimulates the GP.[161] In a bad flier as is *Periplaneta americana*, after the first 2 min of flying the concentration of hemolymph trehalose rose up to 10%, but 5 min after cessation of the flight activity, it reached its initial level. Regarding the octopamine, its concentration decreased not only during the flying, but also after the insect landing. Besides, an increase in the hemolymph volume amounting up to 30% was accompanied by an increase in the concentration of potassium, calcium, and magnesium cations. In some experiments it was established that the presence of octopamine during the takeoff was very important since it affects the rate of carbohydrate oxidation.[138]

During the first phase of flying the carbohydrates were used as fuel, but when the insects continued flying they utilized lipids stored in the fat body. So, after 30 min of the flight activity the concentration of lipids was fourfold increased. The mobilized fat body lipids were transported by the hemolymph as diacylglycerols.[19,75] Since diacylglycerols are water insoluble their transport to the muscles is possible only in the presence of special carriers, the lipoproteins. In the flight muscles the diacylglycerols are hydrolyzed by the enzyme lipoproteinlipase.[138]

In flying *L. migratoria* the CC release the AKH.[162-164] Here, too, the octopamine is responsible for the first rise in the hemolymph lipid concentration,[100] but if the insects will continue flying, the lipid mobilization is governed by the AKH.

In *Schistocerca gregaria* glycerol produced after the hydrolysis of diacylglycerol is utilized as fuel for flying, during which its amount rose, while after the insect landing, it decreased.[165]

Rhodnius prolixus, deprived of the carbohydrate reserves stored in the fat body, utilizes for energy generation during the first 30 min of flight activity triacylglycerol, i.e., the lipids.

Certain flying insects were able to utilize some amino acids, particularly proline, for energy generation. In the flight muscles of flying *Glossina morsitans*, about 20% of proline is oxidized while the rest is recycled. Proline is resynthesized from alanine by the acetyl-CoA during the degradation of the lipids. Proline represents an advantageous fuel since it is recycled from the lipids, the latter being very convenient for storage. Besides, proline and lipids represent high energy sources. During takeoff of *Leptinotarsa decemlineata* proline was utilized as a fuel.

C. HANDLING

In some studies it was evidenced that handling, comprising different stressors such as anesthesia, hanging, water injection, forced locomotion, etc., frequently provoked a rise in the level of blood metabolites (sugars and lipids), leading to hyperglycemia and hyperlipemia. In *P. americana* excitation causes a hypertrehalosemia[159,166] evoked by the release of octopamine.[100,167] Also, in *Locusta migratoria* handling has elicited an elevation of the hemolymph lipid concentration presumably under the effect of octopamine. In *S. gregaria* a hyperlipemia was evoked by temperature and chemical stressors.[101] The handling has induced in *Acheta domesticus* an increase in the levels of hemolymph sugars and lipids. During the initial phase of the stress, the trehalose level rose under the effect of octopamine. Later on, the high trehalose level was replaced by a high lipid level which was provoked by the AKH, whose concentration was elevated under the effect of octopamine.[168] Some researchers have pointed out an interdependence between the reduced reserve materials stored in the insect fat body and the reduced capacity of the insect to mobilize sugars during the stress, which is directly connected with the processes of development.

REFERENCES

1. **Ivanović, J. P. and Marinković, D.,** Individual variability of the pH optimum of amylase in larval populations of the insect species *Morimus funereus* L, *Genetika*, 2, 105, 1970.
2. **Andjelković, M., Milanović, M., and Stamenković-Radak, M.,** Adaptive significance of amylase polymorphism in *Drosophila*. I. The geographical pattern of allozyme polymorphism in the amylase locus in *Drosophila subobscura*, *Genetica*, 74, 161, 1987.
3. **Heinrich, B.,** Ecological and evolutionary perspectives, *J. Insect Thermoregul.*, p. 236, 1981.
4. **Precht, H., Christophersen, J., Hensel, J., and Larchel, W.,** *Temperature and Life*, Springer-Verlag, Heidelberg, 1973.
5. **Hochachka, P. W. and Somero, G. N.,** *Biochemical Adaptation*, Princeton University Press, Princeton, NJ, 1984.
6. **Prosser, C. L.,** *Adaptational Biology: Molecules to Organisms*, John Wiley & Sons, New York, 1986.
7. **Réaumur, R. A.,** *Memoires pour Servir à l'Histoire des Insects*, Vol. 2, d'Imprimerie Royal, Paris, 1736, 1.
8. **Ušatinskaja, R. S.,** *Osnovy holodostojkosti nasekomyh, Izd. Akad. Nauk SSSR*, Moskva, 1957 (in Russian).
9. **Salt, R. W.,** The influence of food on cold hardiness of insects, *Can. Entomol.*, 85, 261, 1953.
10. **Lozina-Lozinskij, L. K.,** The resistance of insects to deep cooling and to intracellular freezing, in *The Cell and Environmental Temperature*, Troshin, Ed., Pergamon Press, Oxford, 1967.
11. **Tanno, K.,** High sugar levels in the solitary bee *Ceratina*, *Low Temp. Sci. Ser. B.*, 22, 51, 1964.
12. **Hanec, W. and Beck, S.,** Cold hardiness in the European corn borer *Ostrinia nubilalis*, *J. Insect Physiol.*, 5, 169, 1960.
13. **Cannon, D. J. and Block, W.,** Adapting to adversity: two distinct strategies in Antarctic microarthropods, in 3rd Symp. on Invertebrate and Plant Cold Hardiness, Cambridge, 1988.
14. **Salt, R. W.,** Principles of insect cold-hardiness, *Annu. Rev. Entomol.*, 6, 55, 1961.
15. **Somme, L.,** Effects of glycerol on cold-hardiness in insects, *Can. J. Zool.*, 42, 87, 1964.
16. **Asahina, E.,** Frost resistance in insects, *Adv. Insect Physiol.*, 6, 1, 1969.
17. **Danks, H. V.,** Modes of seasonal adaptation in insects. I. Winter survival, *Can. Entomol.*, 110, 1168, 1978.
18. **Ring, R. A.,** Insects and their cells, in *Low Temperature Preservation in Medicine and Biology*, Ashwood-Smith, M. J. and Farrant, J., Eds., Pitman Medical, Tenbridge Wells, 1980, 187.

19. **Baust, J. G.**, Mechanisms of cryoprotection in freezing tolerant animal system, *Cryobiology*, 10(3), 197, 1973.
20. **Somme, L.**, Further observations on glycerol and cold hardiness in insects, *Can. J. Zool.*, 43, 765, 1965.
21. **Morrissey, R. E. and Baust, J. G.**, The ontogeny of cold tolerance in the gall fly, *Eurosta solidaginis*, *J. Insect Phy. iol.*, 22, 431, 1976.
22. **Storey, K. B. and Storey, J. M.**, Freeze tolerance in animals, *Physiol. Rev.*, 68, 28, 1988.
23. **Asahina, E. and Tanno, A.**, A large amount of trehalose in a frost resistant insect, *Nature*, 204, 122, 1964.
24. **Rojas, R. P., Lee, R. E., Luu, T. A., and Baust, J. G.**, Temperature dependence-independence of antifreeze turnover in *Eurosta solidaginis*, *J. Insect Physiol.*, 29, 865, 1983.
25. **Grubor-Lajšić, G.**, Biohemijske osnove otpornosti insekata na hladnoću — *Ostrinia nubilalis*, Doctoral thesis, University of Novi Sad, Novi Sad, Yugoslavia, 1984.
26. **Ivanović, J. P., Janković-Hladni, M., Nenadović, V., Frušić, M., and Stanić, V.**, Endocrinological and biochemical aspects of stress and adaptations of phytophagous insects to environmental changes, in *Endocrinological Frontiers in Physiological Ecology*, Sehnal, F., Zabza, A., and Delinger, D. L., Eds., Wroclaw Technical Press, Wroclaw. 1988, 141.
27. **Block, W. and Zettel, J.**, Cold hardiness of some alpine *Collembola*, *Ecol. Entomol.*, 5, 1, 1980.
28. **Ring, R. A. and Tesar, D.**, Adaptations to cold in Canadian Arctic insects, *Cryobiology*, 18, 199, 1981.
29. **Baust, J. G.**, Insect cold hardiness: freezing tolerance and avoidance—the *Eurosta* model, in *Living, in the Cold: Physiological and Biochemical Adaptations*, Heller, B. C., Masacchia, X. J., and Wang, L. C. H., Eds., Elsevier, Amsterdam, 1986.
30. **Ševaljević, L. J., Ivanović-Matić, S., Petrović, M., Glibetić, M., Pantelić, D., and Poznanović, G.**, Regulation of plasma acute phase protein and albumin levels in the liver of scalded rats, *Biochem. J.*, 258, 663, 1989.
31. **Lea, A. O. and Van Handel, E.**, Suppression of glycogen synthesis in the mosquito by a hormone from the medial neurosecretory cells, *J. Insect Physiol.*, 16, 310, 1970.
32. **Thomsen, E.**, Functional significance of the neurosecretory brain cells and the corpus cardiacum in the female blowfly *Calliphora erythrocephala* Meig, *J. Exp. Biol.*, 26, 137, 1952.
33. **Goldsworthy, G. J.**, The effects of removal of the cerebral neurosecretory cells on haemolymph and tissue carbohydrate in *Locusta migratoria migratoriodes* R and F, *J. Endocrinol.*, 50, 237, 1971.
34. **Chino, H.**, Carbohydrate metabolism in diapause egg of the silkworm, *Bombyx mori*. I. Diapause and the change of glycogen content, *Embryologia*, 3, 295, 1957.
35. **Yamashita, O. and Hasegawa, K.**, Studies on the mode of action of the diapause hormone in the silkworm, *Bombyx mori* L. IV. Effect of diapause hormone on the glycogen content in ovaries and the blood sugar level of silkworm pupae, *J. Sericult Sci.*, 33, 407, 1964.
36. **Janda, V. and Slama, K.**, Uber den einfluss von Hormonen auf den Glykogen-, Fett- und Stickstoff-metabolismus bei den Imagines von *Pyrrhocoris apterus* L. *(Hemiptera)*, *Zool. Jahrb. (Physiol.)*, 71, 345, 1965.
37. **Hamilton, M. D., Rojas, R. R., and Baust, J. G.**, Juvenile hormone modulation of cryoprotectant synthesis in *Eurosta solidaginis* by a component of the endocrine system, *J. Insect Physiol.*, 32, 971, 1986.
38. **Tsumuki, H. and Kanehisa, K.**, Effect of JH and ecdysone on glycerol and carbohydrate contents in diapausing larvae of the rice stem borer, *Chilo suppressalis* Walker *(Lepidoptera, Pyralidae)*, *Appl. Entomol. Zool.*, 16, 7, 1981.
39. **Nordin, J. H., Cui, Z., and Yin, C.-M.**, Cold-induced glycerol accumulation by *Ostrinia nubilalis* larvae is developmentally regulated, *J. Insect Physiol.*, 30, 563, 1984.
40. **Hayakawa, Y. and Chino, H.**, Temperature-interconversion between glycogen and trehalose in diapausing pupae of *Philosamia, cynthia ricini* and *pyreri*, *Insect Biochem.*, 11, 41, 1981.
41. **Storey, J. M. and Storey, K.**, Winter survival of the gall fly larvae, *Eurosta solidaginis*: profiles of fuel reserves and cryo protectants in a natural population, *J. Insect Physiol.*, 32, 549, 1986.
42. **Zigler, R.**, Hyperglycaemic factor from the corpora cardiaca of *Manduca sexta*, *Gen. Comp. Endocrinol.*, 139, 350, 1979.
43. **Hayakava, Y.**, Activation mechanism of insect fat body phosphorylase by cold, *Insect Biochem.*, 15, 123, 1985.
44. **Ziegler, R., Hoff, R., and Rhode, M.**, Storage site of glycogen phosphorylase activating hormone in larvae of *Manduca sexta*, *J. Insect Physiol.*, 34(2), 143, 1988.
45. **Copenhaver, P. F. and Truman, J. W.**, Metamorphosis of the cerebral neuroendocrine system in the moth *Manduca sexta*, *J. Comp. Neurol.*, 249, 186, 1986.
46. **Gäde, G. and Spring, J.**, Activation of fat body glycogen phosphorylase in the eastern lubber grossoper *(Romalea microptera)* by the endogenous neuropeptides Ro I and Ro II, *J. Exp. Zool.*, 250, 140, 1989.
47. **Ivanović, J., Janković-Hladni, M., and Frušić, M.**, unpublished data, 1988.
48. **Baust, J. G. and Edwards, J. S.**, Mechanisms of freezing tolerance in Antarctic midge *Belgica antarctica*, *Physiol. Entomol.*, 4, 1, 1979.

49. **Storey, K. B. and Storey, J. M.,** Biochemical strategies of overwintering in the gall fly larva, *Eurosta solidaginis:* effect of low temperature acclimation on the activities of enzymes of intermediary metabolism, *J. Comp. Physiol.,* 144, 191, 1981.

50. **Ramsay, J. A.,** The rectal complex of the mealworm *Tenebrio molitor* L. *(Coleoptera, Tenebrionidae),* *Philos. Trans. R. Soc. London Ser. B,* 248, 279, 1964.

51. **Duman, J. G.,** Insect antifreezes and ice-nucleating agents, *Cryobiology,* 19, 613, 1982.

52. **Duman, J. G.,** The role of macromolecular antifreeze in the darkling beetle *Meracantha contracta, J. Comp. Physiol.,* 115, 279, 1977.

53. **Duman, J. G.,** Environmental effects of antifreeze levels in larvae of darkling beetle, *Meracantha contracta, J. Exp. Zool.,* 201, 333, 1977.

54. **Horwath, K. L. and Duman, J. G.,** Induction of antifreeze protein production by juvenile hormone in larvae of the beetle, *Dendroides canadensis, J. Comp. Physiol.,* 151, 233, 1983.

55. **Zachariassen, K. E. and Hammel, H. T.,** Nucleating agents in the haemolymph of insects tolerant to freezing, *Nature (London),* 262, 285, 1976.

56. **Baust, J. G. and Zachariassen, K. E.,** Seasonally active cell matrix associated ice nucleators in an insect, *Cryoletters,* 4, 65, 1983.

57. **Neven, L. G., Duman, J. G., Beals, J. M., and Castellino, F. J.,** Overwintering adaptations of the stag beetle, *Ceruchus piceus:* removal of ice nucleators in the winter to promote supercooling, *J. Comp. Physiol.,* 156, 707, 1986.

58. **Miller, L. K.,** Cold hardiness strategies of some adult and immature insects overwintering in Alaska, *Comp. Biochem. Physiol.,* 73, 595, 1982.

59. **Stanić, V.,** unpublished results, 1986.

60. **Duman, J. G., Neven, L. G., Beals, J. M., Olson, K. R., and Castellino, F. J.,** Freeze tolerance adaptations, including haemolymph protein and lipoprotein nucleators, in the larvae of the cranefly *Tipula trivittata, J. Insect Physiol.,* 31, 1, 1985.

61. **Sidell, B.,** Cellular acclimatisation to environmental change by quantitative alterations in enzymes and organelles, in *Cellular Acclimatisation to Environmental Change,* Cosins, A. and Sheterine, P., Eds., Cambridge University Press, Cambridge, 1983, 103.

62. **Ballantyne, J. A. and Chamberlin, M. E.,** Adaptation and evolution of mitochondria: osmotic and ionic considerations, *Can. J. Zool.,* 66, 1028, 1988.

63. **White, F. N. and Somero, G. N.,** Acid-base regulation and phospholipid adaptations to temperature. Time courses and physiological significance of modifying the milieu for protein function, *Physiol. Rev.,* 62, 40, 1982.

64. **Ivanović, J., Janković-Hladni, M., Milanović, M., Stanić, W., and Božidarac, M.,** Possible role of neurohormones in the process of acclimatization and acclimation in *Morimus funereus* (Insecta). II. Changes in the heamolymph free amino acid content and digestive enzyme activity, *Comp. Biochem. Physiol.,* 4, 695, 1982.

65. **Baust, J. G. and Miller, K. L.,** Influence of low temperature acclimation on cold hardiness in *Pterostichus brevicornis, J. Insect Physiol.,* 18, 1935, 1972.

66. **Steele, J. E.,** Control of metabolic processes, in *Comprehensive Insect Physiology, Biochemistry* and *Pharmacology,* Vol. 4, Kerkut, G. A. and Gilbert, L. I., Eds., Pergamon Press, Oxford, 1985, 99.

67. **Hoffmann, K. H.,** Der einfluss der temperatur auf die chemische Zusammensetzung von Grillen *(Gryllus, Orthopt), Oecologia (Berlin),* 13, 147, 1973.

68. **Danks, S. M. and Tribe, M. A.,** Biochemical changes in blowfly flight muscle mitochondria following temperature acclimation, *J. Therm. Biol.,* 4, 183, 1979.

69. **Downer, R. G. and Kallapur, V. L.,** Temperature-induced changes in lipid composition and transition temperature of flight muscle mitochondria of *Schistocerca gregaria, J. Therm. Biol.,* 6, 189, 1981.

70. **Singh, S. P. and Das, A. B.,** Biochemical changes in tissue composition of *Periplaneta americana* (Linn.) acclimated to high and low temperatures, *J. Therm. Biol.,* 5, 211, 1980.

71. **Ivanović, J. P., Janković-Hladni, M., Stanić, V., Nenadović, V., and Frušić, M.,** The role of neurosecretion and metabolism in development of an oligophagous feeding habit in *Morimus funereus* larvae *(Col. Cerambycidae), Comp. Biochem. Physiol. A,* 94, 167, 1989.

72. **Ivanović, J. and Djordjević, S.,** unpublished data, 1989.

73. **Friedman, S.,** Carbohydrate metabolism, in *Comprehensive Insect Physiology, Biochemistry and Pharmacology,* Vol. 2, Kerkut, G. A. and Gilbert, L. I., Eds., Pergamon Press, Oxford, 1985, 44.

74. **Hayakawa, Y. and Chino, H.,** Temperature-dependent activation or inactivation of glycogen phosphorylase and synthase of fat body of the silkworm *Philosamia cynthia:* the possible mechanism of the temperature-dependent interconversion between glycogen and trehalose, *Insect Biochem.,* 12, 361, 1982.

75. **Ivanović, J., Janković-Hladni, M., Stanić, V., Nenadović, V., and Milanović, M.,** Midgut of coleopteran larvae the possible target organ for the action of neurohormones, in *Neurosecretion and Neuroendocrine Activity—Evolution, Structure and Function,* Bargmann, W., Oksche, A., Polenov, A. L., and Scharrer, B., Eds., Springer-Verlag, Berlin, 1978, 373.

76. **Applebaum, S. W., Janković, M., Grozdanović, J., and Marinković, D.,** Compensation for temperature in the digestive metabolism of *Tenebrio molitor* larvae, *Phsyiol. Zool.,* 37, 90, 1964.

77. **Janković-Hladni, M. I., Stanić, V., and Ivanović, J. P.,** The effect of temperature and crowding on midgut amylolytic activity in *Tenebrio molitor* adults, *J. Insect Physiol.,* 22, 851, 1976.

78. **Das, A. K. and Das, A. B.,** Compensation for temperature in the activities of digestive enzymes of *Periplaneta americana, Comp. Biochem. Physiol. A,* 71, 255, 1982.

79. **Ivanović, J., Janković-Hladni, M., and Milanović, M.,** Possible role of neurosecretory cells: type A in response of *Morimus funereus* larvae to the effect of temperature, *J. Therm. Biol.,* 1, 53, 1975.

80. **Janković-Hladni, M., Ivanović, J., and Frušić, M.,** unpublished data, 1987.

81. **Truman, J. W.,** Physiology of insect rhythms. I. Circadian organization of the endocrine events underlying the moulting cycle of larval tobacco hornworms, *J. Exp. Biol.,* 57, 805, 1972.

82. **Fuishita, M. and Ishizaki, H.,** Circadian clock and prothoracicotropic hormone secretion in relation to the larval ecdysis rhythm of the saturniid *Samia cynthica ricini, J. Insect Physiol.,* 27, 121, 1981.

83. **Hoffmann, K., Behrens, W., and Ressin, W.,** Effects of a daily temperature cycle on ecdysteroid and cyclic nucleotide titres in adult female crickets *Gryllus bimaculatus, Physiol. Entomol.,* 6, 375, 1981.

84. **Nowosielski, J. and Patton, R.,** Daily fluctuation in the blood sugar concentration of the hous cricket, *Cryllus domesticus, L., Science,* Vol. 144, 3615, 180, 1964.

85. **Gersch, M.,** Kontrolle der circadianen Rhytmik des Trehalosegehaltes der Haemolymphe von *Periplaneta americana, Zool. Jarhb. Physiol.,* 80, 1, 1976.

86. **Johnson, C., Roeber, J., and Hastings, J.,** Circadian changes in enzyme concentration account for rhythm of enzyme activity in *Gonyaulax, Science,* 223, 1428, 1984.

87. **Milanović, M.,** Prečišćavanje crevne alfa amilaze larvi *Morimus funereus.* I Biohemijska karakterizacija poluprečišćenih enzimskih prepara, Magistarski rad, Univerzitet u Beogradu, 1977.

88. **Mutchmor, J. A.,** Temperature adaptation in insects, in *Molecular Mechanisms of Temperature Adaptation,* Prosser, C., Ed., American Association for the Advancement of Science, Washington, D. C., 1967, 165.

89. **Piccone, W. and Baust, J. G.,** The effects of low temperature acclimation of neural (sodium-potassium dependent) ATPase in *Periplaneta americana, Insect Biochem.,* 7, 185, 1977.

90. **Marek, M.,** On the mechanism of some antimetabolic action on the biosynthesis of the "cooling protein" by pupae *Galleria mellonella, Comp. Biochem. Physiol.,* 35, 737, 1970.

91. **Mills, R. R. and Cochran, D. G.,** Adenosinetriphosphatase from thoracic miscle mitochondria of the American cockroach, *Comp. Biochem. Physiol.,* 20, 919, 1967.

92. **Terriere, L.,** The induction of detoxifying enzymes in insects, *J. Agric. Food Chem.,* 22, 366, 1974.

93. **Ivanoviác, J., Naković-Hladni, M., Stanić, V., Milanović, M., and Nenadović, V.,** Possible role of neurohormones in the process of acclimatization and acclimation in *Morimus funereus* larvae (Insecta). I. Changes in the neuroendocrine system and target organs (midgut, haemolymph) during the annual cycle, *Comp. Biochem. Physiol. A,* 63, 95, 1979.

94. **Weeda, E.,** Hormonal regulation of proline synthesis and glucose release in fat body of the Colorado potato beetle *Leptinotarsa decemlineata, J. Insect Physiol.,* 27, 411, 1981.

95. **Mansingh, A.,** Changes in the free amino acids of the hemolymph of *Antheraea pernyi* during induction and temination of diapausa, *J. Insect Physiol.,* 13, 1645, 1967.

96. **Morgan, T. D. and Chippendale, G. M.,** Free amino acids of the haemolymph of the southwestern corn borer in relation to their diapause, *J. Insect Physiol.,* 29, 735, 1983.

97. **Hansen, T., Viik, M., and Luik, A.,** Biochemical changes and cold-hardiness in hibernating beetles, *Ips typographus* (Coleoptera, Ipidae), *Rev. Entomol. (U.S.S.R.),* 59, 249, 1980 (in Russian).

98. **Rudolph, A. S., Crowe, J. H., and Crowe, L. M.,** Effects of three stabilizing agents—prolin, betaine, and trehalose—on membrane phospholipids, *Arch. Biochem. Biophys.,* 245, 134, 1986.

99. **Somme, L.,** The effect of temperature and anoxia on haemolymph composition and supercooling in three overwintering insects, *J. Insect Physiol.,* 13, 805, 1967.

100. **Orchard, I. and Loughton, B. G.,** Is octopamine a transmitter mediating hormone release in insects, *J. Neurobiol.,* 12, 143, 1981.

101. **Davenport, A. P. and Evens, P. D.,** Stress-induced changes in the octopamine levels of insect hemolymph, *Insect Biochem.,* 14, 135, 1984.

102. **Goldsworthy, G. J., Mallison, K., and Wheeler, C. H.,** The relative potencies of two known locust adipokinetic hormones, *J. Insect Physiol.,* 32, 95, 1986.

103. **Robertson, H. A. and Steele, J. E.,** Activation of insect nerve cord phosphorylase by octopamine and adenosine 3'5'-monophosphate, *J. Neurochem.,* 19, 1603, 1972.

104. **Clarke, K. U.,** Histological changes in the endocrine system of *Locusta migratoria* L. associated with the growth of the adult under different temperature regimes, *J. Insect Physiol.,* 12, 163, 1966.

105. **Nanda, D., Ghosla, M. S., and Naskars, S.,** The effects of hypothermia and subsequent rewarming on the neurosecretory neuron in the brain of *Periplaneta americana, Folia Biol. (PRL),* 25, 213, 1977.

106. **Panov, A.,** Reakcija na golodanie A₁ neurosekretornih kletok tutovogo šelkoprjada, *Dokl. Akad. Nauk Sci. SSSR,* 176, 195, 1967.

107. **Ivanović, J. P., Janković-Hladni, M., Stanić, V., and Milanović, M.,** The role of the cerebral neurosecretory system of *Morimus funereus* larvae (Insecta) in thermal stress, *Bull. T. 72nd Acad. Serbe Sci. Arts Classe Sci. Nat. Math. Sci. Nat.,* 20, 91, 1980.

108. **Ishizaki, H., Mizoguchi, A., Hatta, M., Suziki, A., Nagasawa, H., Kataoka, H., Isogai, A., Tamura, S., Fujino, M., and Kitada, C.,** Prochoracicotropic hormone (PTTH) of the silk moth, *Bombyx mori:* 4K-PTTH, in *Molecular Entomology,* Alan R. Liss, New York, 1987, 119.

109. **Agui, N., Granger, N. A., Gilbert, L. I., and Bollenbacher, W. E.,** Cellular localization of the insect prothoracicotropic hormone: in vitro assay of a single neurosecretory cell, *Proc. Natl. Acad. Sci. U.S.A.,* 76, 5694, 1979.

110. **Schmid, A.,** How to use Heidenhains AZA staining in insects, *Neurosci. Lett.,* 101, 35, 1989.

111. **Fournier, B. and Girardie, J.,** A new function for the locust neuroparsins: stimulation water reabsorption, *J. Insect Physiol.,* 34, 309, 1988.

112. **Girardie, J., Faddoul, A., and Girardie, A.,** Characterization of three neurosecretory proteins from the A median neurosecretora cells of *Locusta migratoria* by coupled chromatografic, electrophoretic and isoelectrofocusing methods, *Insect Biochem.,* 15, 85, 1985.

113. **Janković-Hladni, M., Ivanović, J., Nenadović, V., and Stanić, V.,** The selective response of the protocerebral neurosecretory cells of the *Cerambyx cerdo* larvae to the effect of different factors, *Comp. Biochem. Physiol. A,* 74, 131, 1983.

114. **Nenadović, V., Janković-Hladni, M., Ivanocić, J., Stanić, V., and Marović, R.,** The effect of the temperature, oil and juvenile hormone on the activity of corpora allata and the activity of digestive enzymes in larvae of *Cerambyx cerdo* L. (Col. Cerambycidae), *Acta Entomol. Jugosl.,* 18, 91, 1982.

115. **Steele, J. E.,** Hormonal control of metabolism in insects, *Adv. Insect Physiol.,* 12, 240, 1976.

116. **Bogus, M. I. and Cymborowski, B.,** Induction of supernumerary moults in *Galleria mellonella*: evidence for an allatotropic function of the brain, *J. Insect Physiol.,* 30, 557, 1984.

117. **Cymborowski, B.,** Effect of cooling stress on endocrine events in *Galleria mellonella*, in *Endocrinological Frontiers in Physiological Insect Ecology*, Sehnal, F., Zabza, A., and Delinger, D. L., Eds., Wroclaw Technical University Press, Wroclaw, 1988, 203.

118. **Bowler, K.,** Heat death and cellular heat injury, *J. Therm. Biol.,* 6, 171, 1981.

119. **Sinensky, M.,** Homeoviscous adaptation—a homeostatic process that regulates the viscosity of membrane lipids in *Escherichia coli, Proc. Natl. Acad. Sci. U.S.A.,* 71, 522, 1974.

120. **Ivanović, J. and Djordjević, S.,** unpublished data, 1989.

121. **Stephanou, G., Alahiotis, S. N., Marmaras, V. J., and Christodoulou, C.,** Heat shock response in *Ceratitis capitata, Comp. Biochem. Physiol. B,* 74, 425, 1983.

122. **Burdon, R. H.,** Heat shock and the heat shock proteins (review article), *Biochem. J.,* 240, 313, 1986.

123. **Ilinskaja, N. B. and Demin, S. Yu.,** Vlijanie temperaturi na morfologiju politennih hromozom ličinok hironomus v prirode, *Citologija,* 29, 86, 1987.

124. **Ivanović, J. P., Janković-Hladni, M., Spasić, V., and Frušić, M.,** Compensatory reactions at the level of digestive enzymes in relation to acclimatization in *Morimus, asper funereus* larvae, *Comp. Biochem. Physiol. A,* 86, 217, 1987.

125. **Jones, D., Jones, G., and Hammock, B. D.,** Growth parameters associated with endocrine events in larval *Trichoplusia ni* (Hübn) and timing of these events with developmental markers, *J. Insect Physiol.,* 27, 779, 1981.

126. **Hammock, B. D.,** Regulation of juvenile hormone titer: degradation, in *Comprehensive Insect Physiology, Biochemistry and Pharmacology,* Vol. 7, Kerkut, G. A. and Gilbert, L. I., Eds., Pergamon Press, Oxford, 1985, 431.

127. **Raushenbach, I. Yu. and Lukashina, N. S.,** Genetic-endocrine regulation of the development of *Drosophila* under extreme environmental conditions. Communication. IV. Viability, hormonal status and activity of JH esterase in *Drosophila virilis* during larval development on poor diet, *Genetica,* 19, 1995, 1983 (in Russian).

128. **Raushenbach, I. Yu., Lukashina, N. S., and Korochkin, L. I.,** Genetico-endocrine regulation of *Drosophila* development under extreme environmental conditions. I. A study of hormonal status and activity of JH esterase in a *D. virilis* stock selected for resistance to high temperature, *Genetica,* 19, 749, 1983 (in Russian).

129. **Cohen, A. and Patana, R.,** Ontogenetic and stress-related changes in hemolymph chemistry of beet armyworm, *Comp. Biochem. Physiol. A,* 71, 193, 1982.

130. **Cook, B. J. and Holman, M. G.,** Peptides and kinins, in *Comprehensive Insect Physiology, Biochemistry and Pharmacology,* Vol. 11, Kerkut, G. A. and Gilbert, L. I., Eds., Pergamon Press, Oxford, 1985, 532.

131. **Glumac, S., Janković-Hladni, M., Ivanović, J., Stanić, V., and Nenadović, V.,** The effect of thermal stress on *Ostrinia nubilalis* Hbn (Lepidoptera, Pyralidae). I. The role of the neurosecretory cells in the survival of diapausing larvae, *J. Therm. Biol.,* 4, 277, 1979.

132. **Orchard, I.,** Octopamine in insects. Neurotransmitter, neurohormone and neuromodulator, *Can. J. Zool.,* 60, 659, 1982.

133. **Rauschenbach, I. Yu., Lukashina, N. S., Maksimovsky, L. F., and Korochkin, L. I.,** Stress-like reaction of Drosophila to adverse environmental factors, *J. Comp. Physiol. B,* 157, 519, 1987.

134. **Bodnaryk, R. P.,** An endocrine-based temperature block of development in the pupa of the Bertha armyworm, *Mamestra configurara* Wilk, *Invertebrate Reprod. Dev.,* 15, 35, 1989.

135. **Wiens, A. and Gilbert, L.,** Regulation of cockroach fat body metabolism by corpus cardiacum in vitro, *Science (N.Y.),* 150, 614, 1967.

136. **Ziegler, R.,** Metabolic energy expenditure and its hormonal regulation, in *Environmental Physiology and Biochemistry of Insects,* Springer-Verlag, Berlin, 1985, 95.

137. **Siegert, K. J., Krippeit, P., and Ziegler, R.,** Regulation of fat body glycogen phosphorylase during starvation in larvae of *Manduca sexta* (Lepidoptera, Sphingidae), *Acta Endokrinol.,* 246, 32, 1982.

138. **Goldsworthy, G. J. and Wheeler, C. H.,** Adipokinetic hormones in locusts, in *Biosynthesis, Metabolism and Mode of Action of Invertebrate Hormones,* Hoffmann, J. and Porchet, M., Eds., Springer-Verlag, Heidelberg, 1984, 126.

139. **Davies, H. C. and Goldsworthy, G. J.,** Studies on the control of gluconeogenesis in *Locusts migratoria, Gen. Comp. Endocrinol.,* 46, 379, 1982.

140. **Moreau, R., Gourdoux, L., Dutrieu, J., and Benkhay, A.,** Hemolymph trehalose and carbohydrates in starved male adult *Locusta migratoria:* possibility of endocrine modification, *Comp. Biochem. Physiol. A,* 78(N3), 481, 1984.

141. **Baud, L. and Pascal, M.,** Etude comparee de l'evolution du glycogene et du trechalose au cours du jeue chez le Lepidoptere *Bombyx mori* L., *Ann. Nutr. Alim.,* 31, 323, 1977.

142. **Neetles, W. C., Parro, Jr., Sharbaugn, O., and Mangum, C.,** Trehalose and their carbohydrates in *Anthonomus grandis Heliothis zea* and *Heliothis* virescens during growth and development, *J. Insect Physiol.,* 17, 657, 1971.

143. **Bauer, P., Rudin, W., and Hecker, H.,** Ultrastructural changes in midgut cells of female *Aedes aegypti,* L. (Insecta, Diptera) after starvation or sugar diet, *Cell Tissue Res.,* 177, 215, 1977.

144. **Terra, W. R. and Ferreira, C.,** The physiological role of the peritrophic membrane and trehalose: digestive enzymes in the midgut and excreta of starved larvae of Rhynchosciara, *J. Insect Physiol.,* 27, 325, 1981.

145. **Janković-Hladni, M., Ivanović, J., Stanić, V., and Milanović, M.,** Effect of different factors on the metabolism of *Morimus funereus* larvae, *Comp. Biochem. Physiol. A,* 77, 351, 1984.

146. **Rosinski, G., Wrzeszcz, A., and Obuchowicz,** Differences in trehalose activity in the intestine of fed and starved larvae of *Tenebrio molitor, Insect Biochem.,* 9, 485, 1979.

147. **Hill, L. and Goldsworthy, G. J.,** Growth, feeding activity, and the utilisation of reserves in larvae of Locusta, *J. Insect Physiol.,* 14, 1085, 1968.

148. **Chen, P. S. and Hadorn, E.,** Vergleichende Untersuchungen über die freien Aminosäuern in der larvalen Hämolymphe von Drosophila, ephestia und Corethra, *Rev. Suisse Zool.,* 61, 437, 1954.

149. **Strong, F. E.,** The effects of nitrogen starvation on the concentration of free amino acids in *Myzus persicae* (Sulcer) (Homoptera), *J. Insect Physiol.,* 10, 519, 1964.

150. **Bosquet, G.,** Haemolymph modifications during starvation in Phylosamia cynthia Walkeri (Ferber). II. Amino-acids and peptides, *Comp. Biochem. Physiol. A,* 58, 377, 1977.

151. **Janković-Hladni, M.,** Comparative Studies of Midgut Epithelium and Digestive Enzymes during Postembryonic Development of *Tenebrio molitor* L., Doctoral thesis, University of Belgrade, 1971 (in Serbian).

152. **Ivanović, J., Janković-Hladni, M., Stanić, V., and Kalafatić, D.,** Differences in the sensitivity of protocerebral neurosecretory cells arising from the effect of different factors in *Morimus funereus* larvae, *Comp. Biochem. Physiol. A,* 80, 107, 1985.

153. **Grossman, M. and Davey, K.,** The effect of feeding on the cerebral neurosecretory system of the adult male tsetse *Glossina austeni* Newst, *Can. J. Zool.,* 56, 1988, 1978.

154. **Perić, V., Ivanović, J., and Janković-Hladni, M.,** unpublished data, 1990.

155. **Bhaskaran, G.,** Regulation of corpus allatum in last instar Manduca sexta larvae, in *Current Topics in Insect Endocrinology and Nutrition,* Bhaskaran, G., Friedman, S., and Rodriguez, J., Eds., Plenum Press, New York, 1981, 53.

156. **Reddy, G., McCaleb, D., and Kumaron, A.,** Tissue distribution of juvenile hormone hydrolytic activity in *Galleria mellonella, Experientia,* 36, 461, 1980.

157. **van Marrewijk, W. J., van Broek, A. T., and Beenakkers, A. M.,** Regulation of glycogenolisis in the locust fat body during flight, *Insect Biochem.,* 10, 675, 1980.

158. **Candy, D. J.,** Hormonal regulation of substrate transport and metabolism, in *Energy Metabolism in Insects,* Downer, R., Ed., Plenum Press, New York, 1981, 19.

159. **Downer, R. G.,** Induction of hypertrehalosemia by exitation in *Periplaneta americana, J. Insect Physiol.,* 25, 59, 1979.

160. **Goosey, M. W. and Candy, D. J.,** The D-octopamine content of the hemolymph of the locust *Schistocerca americana gregaria* and its elevation during flight, *Insect Biochem.,* 10, 393, 1980.

161. **van Marrewijk, W. J., van der Broek, A. Th., and Beenakkers, A. M.,** Hormonal control of fat-body glycogen mobilization for locust flight, *Gen. Comp. Endocrinol.,* 12, 360, 1986.

162. **Bailey, E.,** Biochemistry of insect flight. II. Fuel supply, in *Insect Biochemistry and Function,* Candy, D. J. and Kilby, B. A., Eds., Chapman and Hall, London, 1975, 89.

163. **Cheeseman, P. and Goldsworthy, G. J.,** The release of adipocinetic hormone during flight and starvation in Locusta, *Gen. Comp. Endocrinol.,* 37, 35, 1979.

164. **Orchard, I. and Lange, A. B.,** The hormonal control of haemolymph lipid during flight in *Locusta migratoria, J. Insect Physiol.,* 29, 639, 1983.

165. **Candy, D., Hall, L., and Spencer, I.,** The metabolism of glycerol in the locust *Schistocerca gregaria* during flight, *J. Insect Physiol.,* 22, 583, 1976.

166. **Matthews, J. R. and Downer, R. G.,** Origin of trehalose in stress induced hyperglycemia in the American cockroach, *Periplaneta americana, Can. J. Zool.,* 52, 1005, 1974.

167. **Downer, R. H.,** Short term hypertrehalosemia induced by octopamin in the American cockroach, Periplaneta americana in *Insect Neurobiology and Pesticide Action,* Neurotox 79, New York, 1980, 335.

168. **Woodring, J. P., McBride, L. A., and Fields, P.,** The role of octopamine in handling and exercise-induced hyperglycemia and hyperlipaemia in *Acheta domesticus, J. Insect Physiol.,* 35, 613, 1989.

Chapter 4

NEUROENDOCRINE SYSTEM IN INSECT STRESS

S. I. Chernysh

TABLE OF CONTENTS

I. INTRODUCTION

In insect endocrinology, priority is traditionally given to investigations on the role of hormones in development and reproduction processes. Meanwhile, a third and very important function of the neuroendocrine system, associated with regulating the organism's adaptation to unfavorable environmental conditions, is still a much less explored area. The only exceptions are mechanisms of diapause, but in this case the protective function of hormones is interpreted as a direct consequence of their morphogenetic activity. At the same time, little research has been done on certain key problems such as the reactivity of the endocrine system in relation to stress effects, its role in regulating intracellular and tissue repair of damage, infection immunity, and catatoxic systems, etc. When, in their reviews, some authors did focus on hormonal responses to stressor action,[1-3] it was mainly pathological aspects of reactivity that were the object of analysis.

The present chapter summarizes experimental data on the multiplicity of responses to stress by the main glands of internal secretion (NSCs of the brain, corpora cardiaca, corpora allata, prothoracic glands); the possible adaptation consequences of these responses are considered, as well as some mechanisms protecting the ontogenetic program from disturbances caused by a stress-induced release of hormones. On this basis, an attempt is made to approach a better understanding of the hormonal mechanism of general adaptation syndrome in insects as compared to that in vertebrates. No doubt, such an attempt today meets with great difficulties since actual evidence is sparse and fragmentary. Therefore the main purpose of this review, besides summarizing experimental data, is not as much to state unquestionable truths as to postulate some hypotheses which stimulate further research.

Another difficulty, besides the paucity of facts, is connected with a certain vagueness of some key concepts, such as "stress" and "general adaptation syndrome". According to the interpretation suggested by Selye,[4,5] the term stress indicates a stereotype state of physiological tension arising in an organism in response to extraordinary stimuli. Besides pathological changes, the state of stress implies a universal complex of adaptive responses defined by Selye as a general adaptation syndrome. However, the assumption that insects, like vertebrates, respond to any kind of damage by a stereotype state of tension or a universal protective reaction is far from indubitable and should be supported by special evidence. Moreover, certain facts point to the existence of various, and in some aspects alternative, forms of nonspecific protective responses to stressor action.

II. NEUROSECRETION

Control of morphogenetic and metabolic processes by the central nervous system (CNS) is mainly mediated by peptide hormones synthesized in the brain neurosecretory cells (NSC). The composition of neuropeptides includes tropic factors regulating the synthesis of ecdysteroids (prothoracicotropic hormones, PTTH) and the juvenile hormone (allatotropic and allatostatic factors), as well as a number of peptides modifying the rate and direction of metabolism in the fat body, Malpighian tubules, and other tissues.

Prior to their release into the hemolymph, the secretory products of cerebral NSCs usually accumulate in the corpora cardiaca or other neurohemal organs. In the glandular part of this organ, a number of hormonally active peptides and biogenic amines of a vertebrate adrenaline type are synthesized, whose secretion is directly governed by the neurons of the brain.

Reorganization of the secretory function of cerebral NSCs under the influence of damaging agents has been recorded in a number of histological studies. Thus, an intensive neurosecretory release from the medial NSCs of the protocerebrum is a widespread response to poisoning with organic insecticides.[6-9] NSC responses to other kinds of damage have been studied less extensively. Antibiotic novoimanin is known to stimulate secret release from medial NSCs in silkworm *Bombyx mori* caterpillars.[10] In *Periplaneta americana* stress stimulates the activity of D cells, leaving A cells passive and inhibiting B and C cells.[11]

The replacement of *Morimus funereus* larvae into an unfavorable thermal condition ($-1°C$) arises in the activity of cerebral NSC of A type.[12,13] Under the heat shock conditions the diapausing caterpillars *Ostrinia nubilalis* show a rapid (during 1 h) activation of synthesis and release of neurosecret from NSC of the A type.[14]

In our experiments[15] a mechanical stimulation of the postdiapausing larvae of *Calliphora vicina* also affected the functional state of NSC (Figure 1). The proportion of cells with a large nucleus, the content of neurosecret in the cytoplasm, as well as the total number of cells which could be identified as NSC significantly surpassed the control level at some stages of stress response. Thus, the increase of peptidergic neurosecretion in certain protocerebrum structures seems to be the usual component of stress response.

Considering the key role of these structures in the insect endocrine system, it can be assumed that stress-induced changes in their activity exert a considerable influence on different regulatory and effector systems directly involved in the process of adaptation.

A. PTTH

The only function of PTTH known at present is stimulation of ecdysone synthesis in the cells of the prothoracic gland or analogous organs. For a long time insects were thought to possess only one form of this hormone. In recent years, however, it has been established that *Lepidoptera* have at least two forms, high-molecular and low-molecular ones. The molecular weights of *B. mori* constitute 4.4 to 5[16,17] and 22 kDa,[18] in *Manduca sexta* 7 and 22 to 28 kDa.[19] The physiological activity of the two forms of PTTH is different. Thus, the low-molecular PTTH form in *B. mori*, which is active in the test with *Samia ricini*, is not competent to stimulate the imaginal development of permanent *B. mori* pupae, whereas the high-molecular form possesses such a property.[18,20] In accordance with the "two PTTH forms—two functions" hypothesis,[21] the high-molecular PTTH form in *M. sexta* insures a sharp rise in the activity of prothoracic glands and the formation of the molting peak of ecdysteroids, whereas the low-molecular form provides the secretion of small amounts of ecdysteroids required for reprogramming hypodermis during the preparation for metamorphosis.

Under normal conditions, PTTH secretion occurs prior to molting, preceding an increase in ecdysteroid synthesis; however, its content in the brain of *B. mori*[22] and *M. sexta*[19] remains relatively constant during the entire period of postembryonic development. On these grounds, PTTH secretion is thought to be regulated at the level of release from neurohemal organs (corpora cardiaca, more rarely corpora allata), and not at the level of synthesis.[19]

Normally, PTTH secretion is regulated by exogenous and endogenous signals, among which three groups of stimuli are distinguished:[19] mechanoreceptive, photoperiodic, and temperature stimuli. Photoperiodic and temperature signals ensure the synchronizing of the life cycle to the climatic rhythm. Mechanoreceptive ones register the attainment of the critical body size and the food input into the digestive tract. Thus, in bugs *Rhodnius prolixus*[23] *Oncopeltus fasciatus*,[24] and in the locust *Locusta migratoria*,[25] stretching of intestine walls serves as a cue to PTTH secretion. Consequently, external signals, including those which may be identified as "stimulation", are able to activate PTTH releasing or synthesis. However, stimulation of mechanoreceptors is known to produce an opposite effect as well. For example, depriving *Galleria mellonella* caterpillars of space required for cocoon formation suppresses the prothoracicotropic activity of the brain and delays pupation.[26]

Direct evidence as to changes in PTTH titer under the influence of stressors has not been obtained so far. However, a number of indirect observations put together allow some tentative conclusions. Thus, information is available of changes in the ecdysteroid titer, the

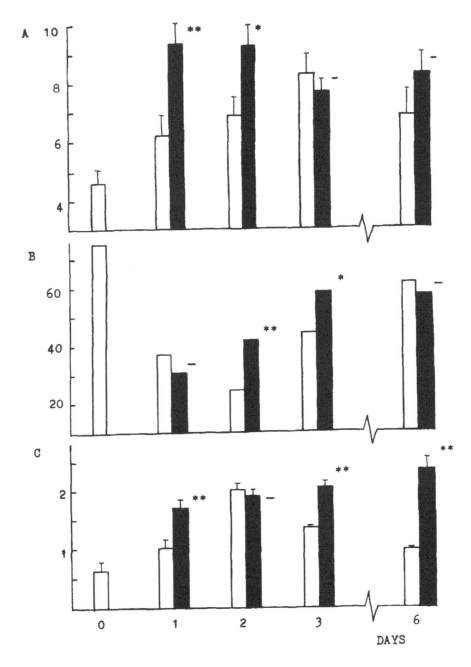

FIGURE 1.   The effect of chronic stress on the functional state of A_2-NSC in the medial group of *Calliphora vicina* protocerebrum. Diapausing larvae were replaced from their hibernation sites ($+3°C$) into $+15°C$ conditions, and their head tips were immediately bound with a cotton thread at a I-II segment level, leaving CNS undamaged. Groups of control (the unshaded column) and stressed larvae, of 20 individuals each, were fixed in Bouin's solution 1, 2, 3, and 6 d after beginning of experiment; paraffined sections of the brain were stained with paraldehyde fuchsin and alcian blue. The number of A_2-NSCs in the medial part of protocerebrum was counted and the areas of their nuclei were measured; the amount of neurosecret in the cytoplasm was assessed by a 4-grade scale. (A) The average number of A_2-NSCs; (B) the proportion of cells with large nuclei (the area of a nucleus section $>/=20 \ \mu m^2$, in % to the total A-NSC number; (C) the amount of neurosecret in the cytoplasm (in conditional units). The significance of the difference from control: − insignificant; * $P < 0.05$; * * $p < 0.01$.

dynamics of the morphogenetic process controlled by ecdysteroids (see Section III.A). Since it is generally accepted that ecdysone secretion from prothoracic glands is directly regulated by PTTH, this information is worth considering.

As seen from the above-cited observations, the stress-induced intensification of secretory function involves the medial protocerebral NSCs in which, according to a widespread opinion, PTTH synthesis is carried out. Hence, it is reasonable to consider these observations as an indirect argument in favor of PTTH involvement in stress response. The force of this argument, however, is somewhat weakened by the fact that medial NSCs can also synthesize other neuropeptides. Moreover, in insects of the *M. sexta* type the source of PTTH may probably be the lateral region of the brain.[27]

More definite evidence for traumatic activation of the prothoracicotropic function of the brain was obtained as result of microsurgical experiments on *G. mellonella* caterpillars.[26,28] In 2- to 3-d-old caterpillars in the final instar, different mechanical damage (cuticle incision, severance of ventral nerve cord, removal of the imaginal disk) causes a precocious molting, and instead of the pupa a supernumerary larval instar is formed.[28] In 4- to 6-day-old caterpillars the same action delays molting. Transplanting the brain from a 2- to 3-d-old individual into a 4- to 6-day-old one induces a precocious molt in the latter. Molting could not be provoked by the juvenile hormone injection: thus, its involvement in this case may be ruled out, and the effects of damage upon 2- to 3-d-old caterpillars may be reasonably attributed to an activation of the cerebral prothoracicotropic function. Under certain conditions, a similar effect is also observed in individuals at the end of their final instar.[26] If at this period caterpillars of *G. mellonella* are deprived of free space required for cocoon formation, the prothoracicotropic activity of the brain is blocked by signals from outer mechanoreceptors. A surgical damage to the integument restores the prothoracicotropic activity of the brain. It should be noted that these two types of stimuli elicit different kinds of response: one of them (limiting in space) delays development, whereas the other (injury) accelerates it. A similar duality of a stressor response has been demonstrated in the response of *C. vicina* (Section III.A); it seems to be a peculiar characteristic of the period preceding metamorphosis. Mechanoreceptive signals which inform the organism of environmental conditions being unfavorable for pupation prolong the stage of a "wandering larva" and thereby give it additional time to find suitable conditions.[29] Damage more dangerous causes an alarm secretion of PTTH and ecdysteroids. Elimination of the morphogenetic block may be a side effect of such a secretion.

B. HYPERTREHALOSEMIC AND HYPERLIPEMIC NEUROPEPTIDES

Besides tropic functions, neurohormones play an important role in regulating the metabolism of lipids and carbohydrates. Mobilization of lipids deposited in the fat body of *Locusta* is carried out by the adipokinetic hormone (AKH), an oligopeptide synthesized mainly by the glandular part of corpora cardiaca.[30] Injected extracts from *Locusta* corpora cardiaca cause hyperlipemia in representatives of other insect orders, which indicates a similarity of AKH in different groups.

AKH function is energy insurance of durable physical loads, particularly the working of wing muscles during flight. In the locust, however, hyperlipemia occurs not only during flight, but also as a result of poisoning with sublethal doses of neurotoxic insecticides.[31-34] Mobilization of lipids is elicited even by a moderate stress associated with handling.[35,36] In the latter case, however, the rapid and short-term rise of diglyceride concentration is due rather to octopamine secretion (see Section II.C), whereas AKH gets involved at the later stages of stress response.

The fall of trehalose concentration in the hemolymph serves as a cue to AKH secretion in the presence of physical load. As a result, the metabolism of lipids appears to be closely associated with that of carbohydrates, which is also controlled by the hormones of corpora cardiaca.

Increase in glycogenolysis and the rise of trehalose concentration in the hemolymph are usual components of stress response.[30] It is probable that the peptide hypertrehalosemic factor of cardiac bodies should play a certain role in this process. However, the available data do not allow any differentiation between its effects and the effects associated with the secretion of biogenic amines.

After the muscles and other organs have gotten additional amounts of lipids and trehalose which they need, the splitting of reserve energy substrates ceases. It is possible that an important role in the metabolism level returning to normal is played by hormones antagonistic to AKH and to the trehalosemic factor.[30] The latter's involvement in stressor response has not been studied, yet it seems probable that in some cases such an involvement should be necessary for homeostasis to be restored and for inadequate hypertrophied responses to chronic stimulation to be prevented.

C. BIOGENIC AMINES

In corpora cardiaca, besides neuropeptides, hormonally active substances are synthesized which belong to the group of biogenic amines similar to vertebrate adrenaline in their structure and physiological activity. Among biogenic amines, in *Orthoptera*[30] and *Lepidoptera*[37] the presence of octopamine has been clearly identified.

Secretion of biogenic amines accompanies the state of stress caused by different stimuli.[36-41] It seems to be one of the most constant and physiologically significant components of insect response to stress.

The literature records mainly pathological consequences of alarm secretion of biogenic amines, studied on the stress-induced paralysis of the cockroach CNS.[1,3] It is doubtless the secretion of these substances, provoked by a prolonged and intensive stimulation, that irreversibly disturbs the activity of the nervous system. Nevertheless, there is reason to suppose that, under less rigid conditions than those leading to CNS paralysis, the stressor secretion off biogenic amines plays an important protective role. One of the facts which is important for understanding the correlation between adaptive and pathological effects of biogenic amines is cited by Sternburg.[1] Electrophysiological investigations show high and low doses of neuroactive substances from corpora cardiaca to produce directly opposite effects upon the functioning of the nervous system: the former suppress and the latter stimulate the spontaneous electrical activity of the ventral nerve cord. Stimulation of nervous activity may in a certain way tell on the speed of behavioral events. This supposition is supported by the fact that *L. migratoria* locomotor activity is stimulated by injection of octopamine, dopamine, or noradrenaline.[42] Thus, the adaptive role of biogenic amines is to some extent reduced to the stimulation of motor responses aimed at avoidance of danger. Such an interpretation accords well with the data on their influence upon carbohydrate and lipid metabolism.

In *L. migratoria*, the concentration of lipids in hemolymph rises already 5 to 10 min after the onset of the insect's mechanical stimulation, and at approximately the same time octopamine secretion grows more intense.[36] It is octopamine which is considered to ensure a rapid mobilization of stored lipids, whereas AKH enters at later periods.[30] In *P. americana*, the injection of physiological doses of octopamine causes hypertrehalosemia as a result of a split in the fat body glycogen.[43] Since mechanical damage affects it in a similar fashion,[44] it is reasonable to assume that biogenic amines are responsible for stressogenic mobilization of carbohydrates as well.[45] As in the case of hyperlipemia observed in *L. migratoria*, an increase in trehalose concentration in *P. americana* hemolymph is recorded not later than a few minutes after stressor action.[44]

Under the conditions of intensive physical load a great need for energy also appears in motoneurons. Biogenic amines ensure that this need should be satisfied, stimulating a split of glycogen stored in the perineurium cells.[30] It is possible that, besides motoneurons, other elements of CNS, such as NSCs, for example, possess catecholaminergic receptors. This may play an especially important role in the development of the *adaptation syndrome*.

Thus, the cerebral neuroendocrine complex responds to stressor action by an urgent ejection into the hemolymph of secretory products which are usually accumulated in corpora cardiaca. Secret composition may include, simultaneously or separately, PTTH, AKH, the hypertrehalosemic factor, biogenic amines, and a number of other substances. In particular, different insects, under different experimental actions, demonstrated a release of the diuretic hormone,[7,46] a cuticle plastification factor,[46] and the neuroactive substance taurin.[47]

Most of secreted neurohormones act directly upon effector organs (motoneurons, muscles, the fat body, the Malpighian tubules), and thus ensure urgent mobilization of resources required for ethological and physiological defensive responses. The very first elements of alarm response are probably formed under the influence of biogenic amines. Further, they are joined or replaced by mechanisms which are controlled by adipokinetic and hypertrehalosemic neuropeptides and seem to be capable of a more prolonged functioning. Changes in PTTH secretion (acceleration, delay, or alteration in qualitative composition of isohormones) cause a reorganization of the activity of the ecdysone-synthesizing system, and thus induce adaptation processes controlled by ecdysteroids.

It is not known in what way, if at all, the set of secreted products changes depending on the species, the stage of ontogeny, or the character of damage. It is possible that, on receiving an appropriate message from the brain, Corpora cardica simply release all hormones into the hemolymph, thereby ensuring a stereotype character of neuroendocrine system response to any stressors. In this case, the possibility of correcting the adaptation syndrome is associated with a secondary neurohormone synthesis at the late stages of stress development. This kind of secondary synthesis was noted in NSCs of *O. nubilalis* under the effect of a heat shock.[14]

III. ECDYSTEROIDS

According to conventional notions, the functions of ecdysteroids are primarily connected with regulation of molting and metamorphosis. In the imago, they control oocyte maturation and the ovulation process. Normally, the dynamics of ecdysteroid titer can essentially differ with various species. However, certain processes are indispensable. In all investigated species ecdysteroids are absent in larvae during the intermolting period or their amount is extremely small. Some hours before molting the titer rises sharply and then as sharply declines. This large peak in the final instar is often preceded by a small one which in *Lepidoptera* causes the cessation of feeding and the transition to the stage of a wandering larva. In insects with complete metamorphosis the ecdysteroid titer is sustained at a high level during the greater part of the pupal period, usually forming two peaks: the first peak induces histolysis of larval tissues, and the second one leads to the formation of imaginal organs. In the imago the titer dynamics reflects the cycle of the sexual glands functioning.[48,49]

A. STRESS-INDUCED CHANGES IN THE TITER

To date no information is available as regards the influence of stressors on ecdysteroid content in the imago. Therefore, the present section will be confined to discussing the responsiveness of preimaginal stages, chiefly those of larvae. Two aspects should be distinguished in analyzing larvae responses: change in the hormone level in the intermolt period and the time when the morphogenetic peak-inducing ecdysis is formed. The effects of damaging agents on the intermolt ecdysteroid titer have been studied by the author on caterpillars of the cabbage moth *Barathra brassicae* and on larvae of *Calliphora vicina*.[50] In intact *B. brassicae* caterpillars in their mid sixth instar, no ecdysteroids were revealed. Six hours after treatment with formaldehyde vapors, estimable quantities of hormones were

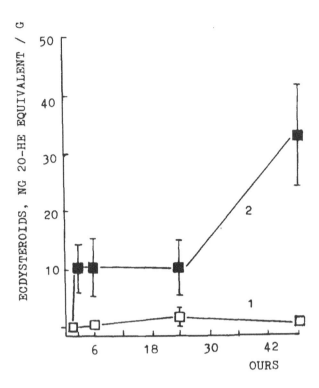

FIGURE 2. Dynamics of the ecdysteroid titer during the chronic
intoxication of *C. vicina* larvae with methanol vapors. The postfeeding
larvae were kept constantly in the chamber with methanol vapors.
Development of stressed and control individuals was retarded by low
temperature (+12°C). The amount of ecdysteroids was measured by
Calliphora biotest method. 1, control; 2, stress.

found by means of biotesting. More detailed experiments were performed with the larvae
of *C. vicina* which had terminated feeding (Figure 2). In the control set, the larvae whose
development had been delayed by a lowered temperature did not practically contain any
ecdysteroids during 48 h of the experiment. Chronic intoxication with methanol vapors
induced synthesis of a small amount of ecdysteroids already 3 h after the onset of the action.
In 48 h the titer reached its maximum, yet even during this period it was much lower than
titer which accompanies *C. vicina* pupariation, according to Karlson and Shaaya.[51] It is not
surprising, therefore, that the stressor stimulation of ecdysteroid synthesis should not be
accompanied by pupariation in this experiment.

However, it would be wrong to think that stress intensifies ecdysteroid secretion in all
cases. In particular, a more complex structure of a response to formaldehyde intoxication
has been recorded for caterpillars of *Bombyx mori*.[52] In intact individuals of fourth instar,
the content of ecdysteroids determined by the Calliphora test method fluctuated at a low
level during the intermolt period, Ten or 12 h prior to the molt onset the titer rose to the
quantity 80 ng/g, whereas during the molt it fell to an indefinable magnitude. In damaged
caterpillars some hours after the damage a slight tendency for surpassing the level in control
individuals was observed, but their morphogenetic peak appeared rather blurred, with a
maximum value 40 ng/g, and the molt itself was delayed.

In analyzing the effects of damaging actions on the dynamics of preimaginal stage
development, it is not difficult to detect considerable differences in insect responses. In
some cases, damage causes a temporary blocking of morphogenesis, while in others, on the
contrary, it accelerates the development. Respectively, sooner or later than the normal time,

ecdysteroids are secreted in morphogenetically effective quantity. These differences may partly be due to the species specificity of response. Systematizing available facts makes it possible to distinguish several reasons for the multiplicity of stress-induced changes in the time span during which the morphogenetic peak of ecdysteroid titer is formed.

One of the frequently observed causes is associated with the appearance of diapause at the larval or pupal stage, which is due to a temporary blocking of PTTH and ecdysteroid secretion. In this respect, changes in *C. vicina*'s responsiveness are characteristic.[53-55] In *C. vicina* the diapause appears at the stage of a postfeeding larva (the prepupa). Under the conditions ensuring larval development without diapause, different damaging agents delay pupariation, whereas in diapausing individuals they produce the opposite effect. Similarly, injection of a sublethal dose of methanol stimulated imaginal differentiation of diapausing pupae of *Barathra brassicae*, whereas damaging fresh ecdysed developing pupae caused in a number of individuals a steady blocking of morphogenesis, lasting for many months and comparable to diapause in duration.[53] Similar differences in the response to mechanical stimulation were revealed in diapausing and developing pupae of *Antheraea jamamai*.[56]

Data on the stimulating influence of stressors upon the development of diapausing larvae and pupae are not confined to the above-cited observations. A number of similar instances can be found in reviews on diapause ecology and physiology.[57,58] An important peculiarity of the response from diapausing insects to stressor actions is the fact that morphogenesis activation starts only after a certain latent period, the duration of which may range from several weeks to many months. An especially prolonged delay of morphogenetic response is exemplified by the pupae of *Pieris brassicae*, whose development is renewed only half a year after integument damage but still somewhat earlier than in the control.[53] A similarly delayed effect has been noted following an injection of 20-hydroxyecdysone to diapausing pupae of *P. brassicae*.[59] On these grounds it can be assumed that the reduction of diapause duration under stress is also due to the subthreshold ecdysteroid secretion which is unable to immediately induce morphogenesis, but reduces the stability of blocking the ''brain-prothoracic glands'' system.

Earlier, the idea of the stimulating influence of injury upon ecdysteroid secretion was formulated on the basis of experiments with diapausing pupae of *Hyalophora cecropia*, in which mechanical damage to integument causes a number of metabolic changes strongly resembling the response to the injection of these hormones.[60-62] The main argument against this hypothesis is the fact that an abdominal fragment of *H. cecropia*, decerebrated and deprived of the prothoracic glands, responds to damage in the same manner as does a whole pupa.[63] To date this objection seems less convincing since it is now known that insects can synthesize ecdysteroids from sterols not only in the prothoracic glands, but also in peripheral tissues.

Anyway, the available facts taken together carry conviction that diapausing insects, as a rule, respond to a stress resulting from damage or stimulation by immediate or delayed activation of ecdysteroid secretion. In developing insects, on the whole, an opposite tendency is characteristic, although exceptions from the rule are known. Let us note, in addition to the above examples, that a delay in molting and ecdysteroid secretion in response to the amputation of outer appendages has been documented for cockroaches[64-66] and crickets.[67] In caterpillars of *Galleria mellonella*, removal of the imaginal disk at the end of their final instar,[28] or stimulation of outer mechanoreceptors by placing the caterpillars in a narrow space,[26,68] delays pupation as a result of blocking the prothoracicotropic function of the brain. However, the removal of the imaginal disk at the beginning of the final instar, on the contrary, leads to precocious molt.[28] It is also known that electrical shock accelerates the development of *Bombyx mori* caterpillars through the ecdysteroid-dependent stages.[69] Thus, we can conclude that developing insects are more various as regards responses to stressor action than are diapausing insects.

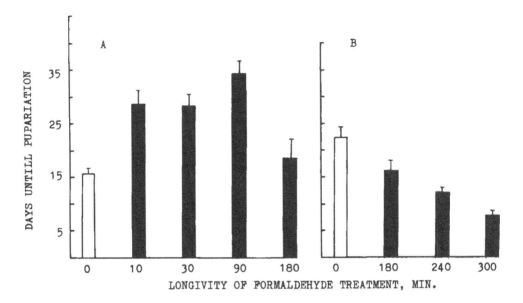

FIGURE 3.   Response of diapausing *C. vicina* larvae to a mild (A) and an intensive (B) treatment with formaldehyde. Diapausing larvae were immersed in a formaldehyde solution, then washed thoroughly and kept at 18°C until pupariation. Untreated larvae (the unshaded column) served as control.

Besides the peculiarities of the physiological status, the qualitative and quantitative characteristics of damage can be a source of the variability of endocrine system response. Thus, in the larvae of *C. vicina* (Figure 3) the period of time required for the passage from the state of diapause to pupariation grew longer during the action of low formaldehyde doses and shortened as a result of more severe damage, as compared to the intact control.[70] It should also be taken into account that every damaging agent, besides producing nonspecific stressor effects, induces pathological processes specific to it. In a number of cases these changes may affect steroidogenesis more strongly than the state of stress per se. Certain agents, such as ionizing radiation, injure endocrine cells selectively, irreversibly suppressing ecdysteroid synthesis.[71-73] It is supposed that regenerating imaginal disks secrete some special inhibitors to suppress the activity of cerebral NSCs and prothoracic gland cells.[74,75] This may, in one way or another, modify the development of stress response.

B. POSSIBLE PROTECTIVE ROLE
The development of insects is regulated by ecdysteroid secretion strictly ordered as to time and level. A sudden change in secretory activity, therefore, may initiate various pathological events associated with the disturbance of the normal course of ontogeny (precocious molt, anomalous metamorphosis) or with a discrepancy between the life cycle and the seasonal rhythm of climate. Nevertheless, there is reason to consider these changes as one of the key elements in regulating the processes of physiological adaptation, since they create conditions required for effective regeneration of damaged structures and functions.

Of special interest in this respect is the rise of ecdysteroid secretion observed in insects with a low level of spontaneous synthesis (larvae in the intermolt period, diapausing individuals, individuals with internal secretion glands removed) when they become subject to stressor action. Our preliminary experiments with 20-hydroxyecdysone injection demonstrated the latter to enhance resistance to pathogenic action.[53] Subsequent investigations have allowed the suggestion that ecdysteroids are stimulators of nonspecific resistance. In particular, Chernysh and Lukhtanov, when experimenting on larvae of *C. vicina* (Table 1), have established that stimulation of resistance is a specific characteristic of low 20-hydroxy-

TABLE 1
Correlation of the Protective and Morphogenetic Effects of 20-Hydroxyecdysone on the C. vicina Larvae

State of larvae	Dose of hormone (ng/individual)	Number of individuals	Time before death (d)	p	Number of individuals with signs of pupariation (%)	p
Ligated developing individuals	Control	193	4.2 ± 0.26	—	2	—
	25	189	4.9 ± 0.30	>0.05	9	<0.01
	50	182	5.7 ± 0.35	<0.001	16	<0.01
	100	164	2.7 ± 0.20	<0.001	59	<0.001
Diapause	Control	42	10.9 ± 1.03	—	0	—
	50	44	16.0 ± 1.85	<0.01	0	>0.05
	100	43	11.4 ± 1.33	>0.05	17	<0.01
	500	43	9.0 ± 0.40	>0.05	78	<0.001

ecdysone doses which cannot induce a molt and morphogenetic events accompanying it. Developing larvae after the feeding period termination were treated with a saturated formaldehyde solution for 120 min; then, by means of ligation the brain, the ring gland, and the synganglion were isolated from the abdominal part of the body and 20-hydroxyecdysone was injected into the abdomen. At a dose of 50 ng per individual, the average time of the abdominal fragment's survival was longer than that of the control, where the larvae were injected with the equivalent quantity of the solvent. A noticeable morphogenetic response to the hormone injection, manifested in the shape of separate spots of brown sclerotized cuticle, was revealed only in 16% of individuals. A higher dose induced an intensive pupariation, the resistance falling lower than the control level. Similar data were obtained from nonligated diapausing larvae (exposure time in formaldehyde being 180 min).

The experiments on C. vicina demonstrate an antagonism between the morphogenetic and the protective functions of ecdysteroids. Concentrations inducing morphogenesis block rather than stimulate the adaptation process. The difference between these two functions is emphasized by the fact that diapausing larvae require much higher doses of 20-hydroxyecdysone to induce pupariation than do developing larvae, whereas protective doses appear approximately equal. Consequently, the fall of sensitivity to morphogenetic stimuli, resulting from diapause formation, does not affect the response to adaptiogenic stimuli. The same relationship between the effects of high and low ecdysteroid concentrations takes place during regeneration. High concentrations, when inducing pupation of Galleria mellonella, suppress the regeneration of wing disks, whereas the presence of a small quantity of ecdysteroids in the hemolymph is a necessary condition of regeneration.[74]

Another aspect of antagonism between the two ecdysteroid effects was revealed in caterpillars, B. mori.[70] 20-Hydroxyecdysone enhanced their resistance to formaldehyde intoxication if it was injected in the middle of last larval instar and did not affect their viability when it was injected not long before cocoon spinning (Figure 4A). A similar result was obtained from experiments on ligated caterpillars with their brain and endocrine glands removed (Figure 4B). Thus, the sensitivity to adaptiogenic 20-hydroxyecdysone stimuli observed in the mid-instar disappears during the preparation for metamorphosis. Target cells, committed to the perception of morphogenetic stimuli, seem to lose their capacity for responding to hormonal actions by the rise of resistance.

A set of experiments on caterpillars of B. mori demonstrates a wide range of the protective actions of exogenous ecdysteroids. Enhancement of caterpillars' resistance to formaldehyde intoxication, whose injurious effect is connected with denaturation of proteins and nucleic

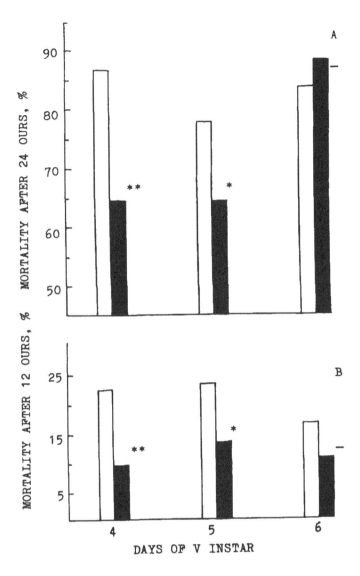

FIGURE 4.   Protective action of 20-hydroxyecdysone on the intact caterpillars of *B. mori* (A) and on those with the brain and prothoracic gland removed (B), on the background of acute formaldehyde intoxication. The caterpillars were subjected to damage (immersion in a saturated formaldehyde solution) on the fourth, fifth, or sixth day after molting. Immediately after that 20-hydroxyecdysone was injected (the shaded column). The control group (the unshaded column) was injected with a solvent. The brain, retrocerebral complex, and prothoracic gland were isolated by ligation at a level of the I abdominal segment.

acids, has been discussed above. In this form of pathology ecdysteroids ensured only a temporary delay of death owing to the rise of tissue resistance, but they could not restore the viability of the organism as a whole.

The case was different when the damage was caused by entobacterin.[76] Entobacterin includes a complex of toxins and spores of the entomopathogenic bacterium *Bacillus thuringiensis*. Poisoning *Lepidoptera* caterpillars with the toxins of *B. thuringiensis* leads to midgut paralysis and to necrosis of intestine epithelium cells, and spore infection results in the development of septicemia. Septicemia causes death of caterpillars later than do toxins,

TABLE 2
Influence of 20-Hydroxyecdysone on the Survival and Body Weight of *B. mori* Caterpillars under the Conditions of Septicemia Epizooty

20-Hydroxyecdysone concentration (mg/l)	Number of individuals	Survival to pupal stage (%)	p	Average mass of cocoon with pupa (g)	p
Control	600	35	—	1.94 ± 0.028	—
0.01	600	45	<0.001	2.01 ± 0.019	<0.05
0.1	600	47	<0.001	2.05 ± 0.025	<0.01
1	1200	37	>0.05	2.09 ± 0.014	<0.001

which makes it possible, using the same model, to appreciate the protective hormone action with respect to poisons and bacterial infection, separately. In the control variant the mortality dynamics in caterpillars treated with entobacterin had two isolated peaks, a toxicogenic one covering the first 2 d and a later peak which, on the basis of the foregoing considerations, may be called "septical". Adding ecdysteroids (a sum of phytoecdysones from the plant *Rhaponthicum carthamoides*, Compositae) to the caterpillars' diet eliminated the first peak and considerably reduced mortality during the subsequent period, thereby demonstrating the effectiveness of defense both against toxicosis and against later pathological effects.

A more direct evidence of antibacterial action of ecdysteroids has been obtained by us against the background of the natural epizooty of septicemia (Table 2). From the beginning of third instar, caterpillars were kept in a place where outbursts of the disease had been observed regularly for previous years. In fourth instar, acute epizooty appeared in the population, with apparent symptoms of bacterial septicemia. At the beginning of fifth instar the survivors received 20-hydroxyecdysone twice in their diet. This proved sufficient for the survival to increase significantly. It was only the highest of concentrations used that turned out ineffective. The general elevation of the culture viability was accompanied by an increase in the average mass of the cocoon. The latter effect has also been noted by Ito et al.[77]

The influence of ecdysteroids on insect immunity to virus diseases can be different. In abdominal fragments of *Bombyx mori* pupae deprived of endocrine glands, replication of the virus of nuclear polyhedrosis appears suppressed as compared to that in normal pupae.[78] Injection of 20-hydroxyecdysone into an abdominal fragment increased the rate of replication, thus demonstrating the dependence of the virus upon the hormonal induction of genome activity. Directly opposite conclusions were made by Keeley and Vinson[79] when experimenting on caterpillars of *Heliothis virescens* infected with the virus of nuclear polyhedrosis. They established that 20-hydroxyecdysone injection delayed the virus replication and, consequently, the death of caterpillars. Similarly, its consumption lowered the frequency of cytoplasmaticpolyhedrosis cases in caterpillars of *B. mori*.[80] In the same caterpillars, 20-hydroxyecdysone was found to exert a stimulating influence on resistance to food stress which provokes the activation of the latent nuclear polyhedrosis virus.[81] There are reasons to consider the indefinitive responses of virus-carrying insects to be due to the duality of physiological functions of ecdysteroids. Intensification of the virus replication in *B. mori* abdominal fragments is caused by morphogenetically effective doses of exogenous hormone.[78] This type of response seems to be immediately associated with the morphogenetic ecdysteroid activity. At the same time, the antivirus action of ecdysteroids on *B. mori* caterpillars is in no way connected with the acceleration of development and morphogenesis. Evidently, the antivirus effect of ecdysteroids is, completely or partially, independent of their morphogenetic activity. It should be specified that enhancement of resistance to a virus disease is not necessarily due to a direct immunological attack against the virus particles. A more important role may be played by a decrease in the organism's sensitivity to stressor actions which provoke the virus transformation from the latent into the pathogenic form.

However, there may be forms of virus diseases which under no conditions can be depressed by ecdysteroid introduction. Thus, it appeared impossible to prevent the development of nuclear polyhedrosis in *B. mori* caterpillars cooled just after the molt to the fifth instar.[81] Among the caterpillars which, after cooling, received various doses of ecdysteroids, the percentage of diseased individuals did not essentially differ from the control level. It should be noted that nuclear polyhedrosis induction when caterpillars are cooled has some peculiar characteristics.[82] Sensitivity to under the low positive temperatures is great only during some early hours after the molt to fourth or fifth instar. Feeding caterpillars before cooling or placing them under starvation conditions after cooling sharply decreases the effectiveness of induction. Evidently, this may be a specific form of induction which cannot be suppressed by nonspecific ecdysteroid-controlled protective mechanisms. The cited fact testifies to the multiplicity of the pathways along which the stressor induction of latent virus occurs. How do ecdysteroids stimulate resistance to various pathogenetic actions? It should be taken into account, as a starting point for our hypothesis, that all the studied ecdysteroid effects are associated with the activation of messenger RNA transcription.[83] The final result depends on which parts of DNA are open to reading-out the information; and thus, in turn, is determined by the accompanying hormonal background and the ontogenic status of target cells. At present there is no direct evidence to prove that an analogous molecular mechanism underlies the adaptiogenic action of ecdysteroids; nevertheless, this supposition seems the most logical and best-argumented one. Keeley and Vinson,[79] in discussing the possible causes of inhibiting the replication in *H. virescens* larvae of nuclear polyhedrosis virus by 20-hydroxyecdysone, assumed this effect to be connected with hormonal stimulation the synthesis of RNA and proteins which are utilized to create the host's new structures and thus appear inaccessible to a parasite. However, the attempts to extend the field of application of this theory meet with considerable difficulties. Thus, the antivirus action of ecdysteroids on *B. mori* caterpillars[80,81] is not accompanied by apparent acceleration of morphogenetic processes which might enhance competition for the products of biosynthesis; at the same time, the hormonal stimulation of the synthesis of RNA and the proteins involved in the imaginal differentiation of the abdominal fragments of *B. mori* pupae results in an intensification of virus replication.[78] It is still more difficult, within the framework of this hypothesis, to explain the protective effects of ecdysteroids in relation to noninfectious damaging factors.

Proceeding from the idea of the interrelation of the protective activity of ecdysteroids and their ability to induce the expression of certain genes, one has to suppose that low, morphogenetically neutral ecdysteroid concentrations stimulate the synthesis of enzymes involved in repairing intracellular, tissue, and organ lesions.[70] The complex of ecdysteroid-induced enzymes also comprises microsomal oxidases of mixed function[84,85] whose presence ensures inactivation of various toxins and thus can be one of the key factors of nonspecific adaptation. A fairly important role in adaptation processes requiring a considerable expenditure of energy may be played by the hormonal regulation of the activity of carbohydrate metabolism. In *B. mori*, 20-hydroxyecdysone accelerates glucose metabolism and changes its direction: in the absence of the hormone, glucose is converted into glycogen and is deposited in the fat body, whereas in its presence it is utilized for the synthesis of the mobile form of carbohydrates—trehalose.[83] It is possible that the genes controlling these and other processes which are important for intensifying the organism's resistance are particularly sensitive to hormonal stimulation and present the main target for low ecdysteroid concentrations.

A temporary blocking of ecdysteroid secretion seems also to be of a certain adaptive significance. Responses of this kind result in a delay of the next molt and a prolongation of the larval or prepupal stage. Positive consequences of morphogenesis delay are most obvious in cases of mechanical damage to outer appendages and the germs of imaginal organs. Since a molt depresses regeneration of either of these,[74,87] its delay increases the time reserve for the regeneration of damaged organ. One should also take into account that a molt is associated with an intensified DNA reduplication in hypodermis and other organs. Damage to the primary DNA structure caused by different agents, be it ultraviolet irradiation, ionizing radiation, or chemical mutagens, is known to be repaired by special enzyme systems as far as the reduplication stage. After a new DNA molecule has been synthesized on the damaged matrix chain, an irreversible mutation takes place. Thus, the adaptive value of a temporary blocking of the morphogenetic ecdysteroid peak may be the creation of additional time for the repair of damage at different levels, from point mutations of DNA to the loss of whole organs. However, in some pathological situations such a type of response may be useless; moreover, it may present an additional danger. Thus, under the conditions of epizooty, individuals retarded in their development undergo a greater risk of being infected with a pathogen whose concentration in a population and its habitat tends to rise rapidly. Due to this, larvae which are capable of responding to a contagious infection by stress-induced acceleration of development may have an advantage over the rest of the population. Consequently, both acceleration and delay of the morphogenetic ecdysteroid peak formation may be advantageous, depending on the particular nature of the pathogen and concomitant factors. Hence, the optimal strategy of response to stressor action should allow an appropriate correction. During mechanical damage, when the expedience of molt delay is especially urgent, the possibility of correction is associated with the fact that the regenerating tissue secretes certain inhibitors which, directly or indirectly, block ecdysteroid synthesis.[74,75]

IV. JUVENILE HORMONES

PTTH, ecdysteroids, and the group of three acyclic sesquiterpenes, defined by the common term "juvenile hormone" (JH), form the classical triad of regulators of insect development. In preimaginal states, JH concentration determines the direction of hypodermis cell differentiation and, consequently, the type of ecdysteroid-induced molt. In the female, JH stimulates vitellogenin synthesis in the fat body and thus contributes to the maturation of gonads. Its targets, besides hypodermis, the fat body, and gonads, are the nervous and neuroendocrine systems. JH neurotropic effects are mainly confined to the rise of general locomotor activity and initiation of specific forms of sexual behavior.[88] The influence of JH upon endocrine gland activity is not simple and leads to directly opposite results (induction or inhibition of hormone release), depending on the species of the insect and the stage of ontogenesis. Implantation of corpora allata activates the prothoracic glands and causes imaginal development of decerebrated pupae of *Hyalophora cecropia*, *Samia cynthia*, and *Bombyx mori*.[89] The same effect is produced upon *H. cecropia* and *Antheraea polyphemus* by JH analogues.[90] On the basis of these facts Krishnakumaran and Schneiderman[90] advanced a hypothesis according to which the prothoracic glands are subject to double control by PTTH and JH. In the intermolt period JH stimulates ecdysteroid secretion in small quantities required for the growth of many tissues, such as imaginal disks which continue growing throughout the larval stage. PTTH periodically superactivates the prothoracic glands, as a result of which ecdysteroids are secreted in quantities sufficient for the induction of molt or metamorphosis. Prothoracicotropic properties of JH are mainly illustrated by the data obtained from *Lepidoptera* pupae. However, this phenomenon is more widely spread. Thus, according to unpublished evidence obtained by Chernysh and Kratina, the application of JH analogue altozid causes a cessation of diapause and the pupariation of *Calliphora vicina* larvae, i.e., processes associated with an accelerated activation of the ecdysteroid synthesis system. At the same time, JH prothoracicotropic effects were not observed in experiments with representatives of *Hemiptera* and *Blattoptera*.[90] There is an impression that the given type of response to JH is characteristic of *Holometabola*, but not of *Hemimetabola*, i.e., it is in a certain way associated with the presence of the pupal stage in the life cycle.

In contrast to the response of prothoracic glands, PTTH secretion in *Lepidoptera* is usually inhibited by the presence of a high JH concentration in the hemolymph. Thus, in *Manduca sexta* caterpillars, PTTH secretion at the end of the final instar occurs only after the titer of JH has lowered.[91,92] In certain species such an effect may be the basis of the mechanism of larval diapause.

The level of JH synthesis is controlled by a complex system of stimulating and inhibiting signals whose relationship varies considerably in different species. Neural regulation is carried out by the neurons of the supraesophageal and the subesophageal ganglia. Neurosecretory material which contains allatotropins and allatohibins, i.e., neuropeptides possessing antagonistic properties, is released via allatal nerves and also humorally from brain NSCs into corpora allata. The peculiarity of governing JH titer is that both the synthesis and the level of metabolic inactivation of the hormone by specific JH-esterases are subject to regulatory influences.[93] The activity of JH-esterases in *Lepidoptera* is enhanced by an unknown factor secreted from the brain or from the subesophageal ganglion.

There are almost no direct data on the effects of stressors upon JH titer. Nevertheless, the number of indirect indications is sufficiently great. Several examples will suffice for illustration. It is known that surgical damage causes supernumerary molts in cockroach *Leucophaea maderae*.[94] This effect is considered to be associated with the prolongation of corpora allata activity. Amputation of antennae in the cockroach *Periplaneta americana* produces an analogous juvenilizing effect.[95] In the butterfly *Choristoneura fumiferana*, virus infection, as well as treatment with JH analogues, leads to the formation of larvoids, which is considered as evidence of the stimulation of JH synthesis by a pathogen.[96] An abnormally high titer has been found in *Spodoptera litura* caterpillars infected with the virus of nuclear polyhedrosis.[97] In *Galleria mellonella*, mechanical damage seems to induce not only prothoracicotropic but also allatotropic brain activity; hence, the transplantation of the brain from damaged caterpillars which are at the beginning of seventh instar to caterpillars at the end of the instar elicits in the latter a supernumerary larval molt.[28] A sharp cooling of caterpillars also induces JH secretion and supernumerary molts.[98]

The cited facts testify that physical and biological injuries often cause a more intensive or more prolonged JH secretion from corpora allata. Pathological consequences of this response are obvious, and find expression in various anomalies of development. Nevertheless, it is relatively seldom that stress results in serious morphogenetic disturbances. Ontogenesis stability in spite of unfavorable external influences is ensured, at least partially, by special mechanisms which will be discussed in the next section. Here it should be reasonable to focus on some hypotheses which deal with possible adaptive consequences of JH secretion during stress.

Analyzing the dynamics of the development and mortality of *Drosophila virilis* larvae under unfavorable environmental conditions, Raushenbach and collaborators[99,100] have come to the conclusion that the principal cause of mortality is the blocking of the pupariation process as a result of inhibiting PTTH and ecdysteroid secretion. Advantage in adaptation was on the side of those individuals in which, under the given conditions, the gene activator of the JH esterase was not expressed. Such individuals were able to overcome the critical stage of ontogenesis owing to preservation in the hemolymph of the JH titer which was sufficient for prothoracic gland activation. Within the framework of this hypothesis the leading role is attributed, not to the rise of JH secretion, but to blocking its metabolical degradation, Yet the result proves to be the same as during the stressor activation of corpora allata. It is remarkable that the JH protective effect in this case is mediated by ecdysteroids. The fact emphasizes the interrelation of separate elements of a common mechanism of adaptation.

The significance of JH as activator of ecdysteroid synthesis is probably more universal and passes beyond the limits set up by the above hypothesis. In the intermolt period the prothoracicotropic effect of JH may contribute to the rise of cell resistance as a result of the activation of ecdysone-dependent macromolecular synthesis. Owing to this, or resulting from a direct action on the enzyme and immunological protective systems, introduction of JH analogues induces the main catatoxic system, i.e., microsomal mixed function oxidases in the housefly *Musca domestica*,[101] and suppresses the development of nuclear polyhedrosis virus in caterpillars of *Hyphantria cunea*.[102]

Stimulation of JH synthesis or delay of JH degradation may cause a rise of resistance in a different way as well. It is known that a high JH titer, at least in some *Lepidoptera* species, often promotes the formation of larval diapause. In such cases, a stress-induced intensification or prolongation of JH secretion may cause the state of diapause. However, it is just diapausing insects that are distinguished for a high nonspecific resistance, partly owing to a low general level of metabolism.[103] The logical chain of such facts leads to the idea that in insects with the above type of diapause regulation a stress-induced rise of the JH titer under certain conditions switches a mechanism of passive defense against damage, which is associated with a temporary blocking of growth, morphogenetic transformations, or reproduction.

V. STRESS AND STABILITY OF ONTOGENETIC PROGRAM

Insect postembryonic development comprises a sequence of stages strictly ordered in time; each of them is induced by thoroughly balanced titers of ecdysteroids and JH. Therefore the involvement of these hormones in stress response creates a potential danger for the normal course of ontogenesis. Irrespective of the fact whether the reorganization of their titers is, or is not, directed toward the rise of the organism's resistance, the neuroendocrine system must possess adequate ensuring mechanisms whose switching-on secures ontogenesis stability during sharp fluctuations of the environment. The few facts that are available to date do not allow an overall consideration of this problem. Nonetheless, it seems reasonable to make some tentative observations and suggest some approaches to its solution.

As shown by experiments on *Calliphora vicina* larvae (Section III.B), enhancement of cell resistance is promoted by a lower ecdysteroid titer than the one that is required to induce pupariation. A quantitative differentiation between the protective and morphogenetic effects, when cell resistance is stimulated by a low hormone concentration, whereas molt and other morphogenetic processes are stimulated by a higher one, serves as one of the prerequisites for normal development in abnormal conditions. On the other hand, numerous data obtained from biotesting of ecdysteroids show that the epidermis sensitivity to morphogenetic stimuli in *Diptera* larvae is supported at a high level only for a few hours prior to the onset of metamorphosis. The nonsensitivity of tissues in the intermolting period should evidently be regarded as still another mechanism ensuring the ontogenetic program.

The same mechanisms may function in cases of alarm JH secretion. However, the general sensitivity of various tissues to JH is sustained during longer periods of ontogenesis, embracing a considerable part of the intermolting period, the pupal and imaginal stages.[104] Consequently, the possibility that stress-provoked JH secretion may disrupt the normal course of ontogenesis is much higher than in the case of urgent secretion of ecdysteroids.

Probably, there may also be a third way of stabilizing the ontogenetic program under stress conditions, which may be connected with the active reparation of damage in target cells. This hypothesis has been formulated by us on the basis of our experiments on the larvae and pupae of *Musca domestica* treated with juvenoid.[105] The treatment of the larvae with juvenoid causes the delayed effect, i.e., a blocking of imago eclosion from the puparium. However, the retardation of pupal development restores the fly's ability of eclosion (Figure 5). The possible interpretation of this fact with consideration of supplementary data[105] is that the growth of the latent period affords additional time for the reparation of functional disturbance in target cells.

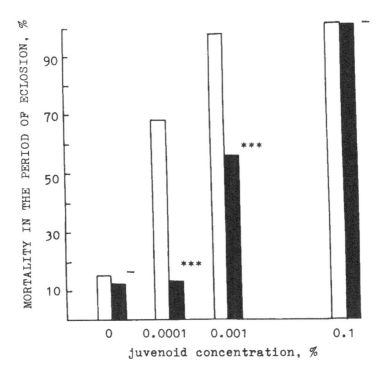

FIGURE 5. Blocking by juvenoid the eclosion of the imago of *Musca domestica* under the conditions of a rapid (unshaded columns) and a delayed metamorphosis. *M. domestica* prepupae were immersed for 1 h in a solution of altozid preparation (Zoecon corporation, the active principle of juvenoid methoprene). Before the imago eclosion one group of individuals was constantly kept at +25°C, whereas the other group was kept at +17°C during a week and subsequently at +25°C. In the second group, as compared to the first one, the duration of pupal development increased by 4 d. The figure presents the data on the mortality of pharate imagous in the puparium.

Nothing is known at present about the protective role of such reparative mechanisms under stress conditions; nor is there any reason to rule out their involvement in stabilizing the morphogenetic process in addition to other adaptations, such as confining the fluctuations of hormone titers within morphogenetically safe limits or lowering the sensitivity to them of target tissues during the intermolt period.

VI. COMPARATIVE ASPECTS OF STRESS ENDOCRINOLOGY

With respect to morphofunctional characteristics, many of the elements in the insect neuroendocrine system surprisingly resemble analogous structures in vertebrates.[106-109] In recent years, using immuno- and physicochemical methods of analysis, over 20 compounds have been found which are identical, or at least very close, to vertebrate hormones. In the brain, the corpora cardiaca and corpora allata of different species, a number of peptides have been revealed, such as peptides with the immunoreactivity of the P-substance, somatostatin, releasing factors of somatotropin and luteotropin, glucagon, insulin, gastrin-cholecystokinin, vasoactive intestinal peptide, pancreatic polypeptide, secretin and endogenic opioids, beta-endorphin, alpha-melanocyte-stimulating hormone, vasopressin, neurophysin, encephalin,[109-111] adrenocorticotropic hormone,[112-114] gonadoliberin, and luteinizing and folliculostimulating hormones.[115] In the body of insects some nonpeptide hormones have also been found, such as sex steroid progesterone and testosterone.[116-118] Biogenic amines (see Section II.C), prostaglandins,[111] serotonin.[119] Thus, the idea that vertebrate and insect hormones have a common origin and both emerged at the earliest stages of evolution[120,121] is supported by an increasing body of facts. At the same time, 600 million years of independent evolution deeply affect the organization of each neuroendocrine system. The asymmetry of hormone composition makes conspicuous the ability of insects for the synthesis of isoprenoid juvenile hormones which have no apparent analogues in the vertebrate endocrine system. Insect ecdysteroids strongly differed in a number of parameters from vertebrate steroid hormones, though certain structural and functional parallels can still be traced.[70]

On the basis of the idea of similarity in the general pattern of the internal secretion organs and the evolutionary relationship of the main hormone groups, it seems promising to compare the stress response mechanism in vertebrates and in insects. Most interesting in this context are the classical "stress" hormones of the vertebrates: biogenic amines, corticosteroids, ACTH.

Biogenic amines, adrenaline and noradrenaline, are synthesized by the neurons of the vertebrate sympatic nervous system and mediate the transmission of the nervous impulse in adrenergic synapses. As hormonal regulation factors, these same substances are formed and secreted into the blood mainly by the chromaffine adrenal tissue which in embryogenesis also develops from the germs of sympathetic ganglia. Cannon[122] was the first to postulate an important role of this group of hormones in an organism's adaptation to the action of strong stimuli; and since then this concept has been generally recognized. The urgent release of biogenic amines during stress stimulates blood circulation, oxygen consumption, splitting of glycogen, and lipids at their storage sites, with a corresponding rise of glucose and fat acids content in the blood plasma. Besides providing energy for the work of peripheral effector systems, biogenic amines stimulate corticosteroid secretion from the adrenal cortex[123] and the formation of the state of enhanced nervous excitability and emotional tension.

In insects, the state of stress is also accompanied by an increased secretion of biogenic amines (see Section II.C). The metabolic effects attained (hyperglycemia and hyperlipemia) correspond to those observed in vertebrates. The ways of regulating the stress-induced release of hormones are also similar. Both in insects and in vertebrates monoamine secretion is subject to direct control of CNS, while the tropic neuropeptides or releasing factors, whose mediation is usual as regards other endocrine glands, are not involved. This ensures a quick response to an extraordinary stimulus, which is critically important for behavioral adaptation. The intensive and prolonged adrenaline secretion conditioned by an excessive emotional tension is at first a certain gain, since it more effectively provides the effector organs with energy substrates and oxygen, but further on it may be fraught, at least for mammals, with serious complications, such as infarction of heart muscle. In insects, the structure of the circulatory and the respiratory systems rules out the possibility for the development of such a pathology: here, the hypersecretion of biogenic amines usually results in an overstrain and paralysis of CNS. However, in both cases the same phenomenon is virtually observed, i.e., the unrealized defensive reaction develops into an adaptation disease.

Thus, with respect to function, the role of biogenic amines in insect stress response is very similar, one might even say identical, to that of vertebrates. Their chemical affinity is also beyond doubt. Hormonally active insect monoamines are represented by octopamine which, in vertebrates, plays the role of an immediate precursor of noradrenaline.

According to the stress conception of Hans Selye,[5,124] the central place in regulating the general adaptation syndrome in vertebrates is held by corticosteroids, particularly glucocorticoids. Intensive glucocorticoid secretion from the adrenal cortex or homologous formations in lower vertebrates stimulates gluconeogenesis at the expense of desamination of amino acids. As a result, against the background of a general shift of protein metabolism toward catabolism, the provision of effector organs with glucose is increased. At the same time, in the liver, the synthesis of protective enzymes such as microsomal mixed function oxidases is induced, which plays an important role in the development of catatoxic defensive responses. Moreover, glucocorticoids produce an antiinflammatory action, depressing the hyperreactivity of the immune system and thus ensuring the organism's passive defense from damage caused by strong stimuli. The combination of syntoxic and catatoxic properties,[125] together with the stimulation of gluconeogenesis, seems to be the essence of the protective role of glucocorticoid secretion during stress.

In insects, glucocorticoids may also be involved in stress response. In our experiments, which were carried out by Chernysh, Priyatkin, and Simonenko, in *Calliphora vicina* hemolymph were substances found immunologically related to cortisol, progesterone, and testosterone. Under the chronic stress conditions induced by ligating the front tip of the bodies of postdiapausal individuals, the concentration of these substances in the hemolymph underwent regular changes (Figure 6). The cortisol content increased progressively, surpassing the control level 1 d after the beginning of stressor action. Such a response resembles the changes in glucocorticoid concentration in vertebrates. The testosterone content in the hemolymph was somewhat lower than normal during the initial period, but by the third day grew higher than the norm. Supposing that in insects testosterone acts as protein synthesis stimulator, as it does in vertebrates, these data agree with the concept of Selye's general adaptation syndrome, according to which the early stages of stress are characterized by catabolism and the later ones by anabolism. Progesterone turned out to be the only hormone whose concentration in the hemolymph was steadily lower than the norm. The physiological meaning of this phenomenon is obscure, but it cannot be ruled out that it is determined by an acceleration of conversion of progesterone to cortisol.

The facts discussed in Section III indicate that insect ecdysteroids perform a function comparable in some respects with that of glucocorticoids. Above all, it concerns their ability to induce the expression of genes responsible for the synthesis of certain groups of protective enzymes, including microsomal mixed function oxidases. Like glucocorticoids, ecdysteroids seem in some cases to shift metabolism toward the generation of mobile carbohydrate forms, owing to which the pool of utilized energy substrates is replenished.[86] However, the mechanism of this effect may essentially differ from what occurs in vertebrates. In any case, in the work cited, the trehalose is considered to be formed of carbohydrates and not of reserve amino acids. Consequently, the ecdysteroid effect is not based on increase in catabolism. Considerable differences seem to characterize the role of these hormones in governing the immune system. While glucocorticoids act as inhibitors of cellular and humoral immunity, suppressing inflammatory reactions and causing thymus involution, ecdysteroids do not produce such a kind of effects. On the contrary, the few facts available testify rather to an activation of antiinfection immune systems under the influence of ecdysteroids. It should be noted in this connection that the organization of the insect immune system renders unlikely the possibility of pathological abnormalities such as allergic or inflammatory diseases of vertebrates; hence there is no need in hormonal inhibition of immunological responses.

The metabolic and possibly evolutionary precursor of ecdysteroids, as well as of corticosteroids, is cholesterol—a structural membrane component in all metazoans. In onto- and phylogenesis, corticosteroids lose the side chain of 6 carbon atoms, while ecdysteroids acquire a higher level of hydroxylation. These and some other kinds of reorganization form considerable differences in their structure. Nevertheless, a curious coincidence attracts attention. The type of physiological corticosteroid activity is determined by the presence of a hydroxil group (or an oxygen atom) in a certain position. Glucocorticoids possess such a group, mineralocorticoids do not; as for the rest, their chemical structures may be quite identical. 20-Hydroxyecdysone differs from ecdysone only by the presence of the hydroxyl group in C-20 position. Particularly essential is the circumstance that in mammals ecdysone causes such effects as are characteristic of mineralocorticoids (see Reference 126), whereas 20-hydroxyecdysone induces enzyme activity in hepatocytes as glucocorticoids (see Reference 127). Besides demonstrating parallelism of structure and function, these facts suggest

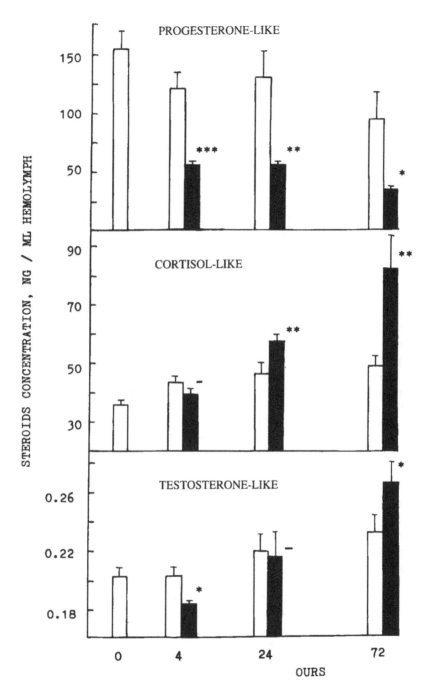

FIGURE 6. The effect of chronic stress on steroid dynamics in the larvae of *C. vicina*. A stimulating ligation was performed during the replacement of diapausing larvae from their hibernation sites ($t°$ +4°C; the hibernation period 2 months) into +17°C conditions. Immediately after replacement and also 4, 24, and 72 h later, hemolymph samples were taken both from the ligated and the control (the unshaded column) individuals (five replications, each of which contained hemolymph samples from about 200 larvae). The samples were subsequently subjected to the standard procedure of radioimmune assay, using specific antiserums against progesterone, cortisol, and testosterone.

the existence in mammals of receptors which are able to perceive the presence of ecdysteroid as a hormonal stimulus. This hypothesis is also supported by many years of experience in the medical application of preparations of the plant *Rhaponthicum carthamoides. R. carthamoides* extraction containing ecdysteroids enhances resistance to stressor actions and to intensive psychic and physical load. On the other hand, hydrocortisone injection did not affect the molt of insects.[53,128] Obviously, corticosteroids are entirely lacking in the morphogenetic activity characteristic of ecdysteroids, since it requires the presence of the side chain and some other elements of structure.

In vertebrates, steroid hormones are combined in four large groups (glucocorticoids, mineralocorticoids, estrogens, and androgens), each of which realizes a coordinate battery of regulative influences. Insects have no possibility of such a differentiation of ecdysteroid functions because ecdysteroids are usually represented by only two hormones, ecdysone often being but a short-lived precursor of 20-hydroxyecdysone. Differentiation of functions which, in vertebrates, is ensured by the diversity of chemical steroid composition, must be achieved in some other ways where ecdysteroids are concerned.

One of such ways may be a more complex two-level regulation of ecdysteroid secretion (see Section III), at which a low hormone titer induces adaptive reorganization of metabolism and a high one induces morphogenesis. In this context, a comparison between the mechanisms of stressor steroidogenesis induction in insects and in vertebrates is of a particular interest. In vertebrates, glucocorticoid secretion by the adrenal cortex is stimulated by the adrenocorticotropic hormone synthesized by adenohypophysis cells in response to stressor excitation of the hypothalamus area. In the neurosecretory cells of the insect brain an ACTH-like substance is also formed.[112-114] Its function is unknown, but ACTH injection into decerebrated diapausing pupae of the *Barathra brassicae* stimulated imaginal development of some of them.[129] This action imitates the effect of PTTH, the main regulator of ecdysteroid synthesis; hence, it may be thought that ACTH-like substance in the insect brain possesses some properties of PTTH. However, the hormones united under the term PTTH are represented by several forms distinguished both by their physicochemical properties and by their physiological activity. Amino-acidic sequence of one of these forms 4KD-PTTH from heads of *Bombyx mori* imagos—is about half coincident with that of vertebrate insulin.[130] Introduction of insulin into intact and decerebrated diapausing pupae of *Pieris brassicae* stimulates ecdysteroid synthesis and imaginal development.[131] Besides relatively low-molecular (4-4, 4KD) insulin-like and ACTH-like peptides, prothoracic glands in *Lepidoptera* are activated by polypeptides with 22 to 27 kDa molecular weight, whose properties have not been studied well enough to allow comparing them with the hormones of the vertebrates. However, even ruling out such a comparison, it is clear that ecdysteroid synthesis is subject to a more complex and comprehensive control on the part of neuropeptides than is glucocorticoid secretion. Some elements of this complicated mechanism may have emerged in the common ancestors of insects and vertebrates; others are new formations in the process of evolution, characteristic of insects and possibly of some groups related to them.

The comparison of the protective role of biogenic amines, on the one hand, and steroid hormones, on the other hand, leads to the conclusion that in the process of evolution of the animal endocrine system the former appeared to be an extremely conservative element, whereas the latter fairly soon diverged. To understand the causes of this difference, the following circumstances should be considered. The function of biogenic amines at the molecular level is mainly reduced to activating the consumption of carbohydrate and lipid energic reserves. However, the main ways of deriving energy contained in carbohydrates and lipids appeared long before the appearance of the first multicellular organisms, and they have not practically changed.[132] Under these conditions there was evidently no need for any essential reconstruction of the system of biogenic amines.

Other regulators of carbohydrate metabolism are also conservative. As mentioned above, insects synthesize insulin-like and glucagon-like peptides which by their immunochemical and physicochemical properties are close to insulin and glucagon of vertebrates. Their metabolic effects are also similar: the insulin-like factor causes hypoglycemia and the glucagon-like one elicits trehalosemia.[133-135] Both in insects and in vertebrates hormone synthesis takes place in the endocrine cells of the intestinal tract.[136] In insects, though, the same or similar factors are also secreted by the central neuroendocrine organs via *corpora cardiaca*. The latter fact allows certain differences in the organization of the insect stress response, since it makes possible a direct and rapid neurogenic stimulation of secretory activity. In vertebrates, pancreatic insulin and glucagon play a less immediate role in stress response, for their secretion is controlled by peripheral feedback mechanisms.

Steroid hormones are much more diversified as to their functions than are biogenic amines. In the given context the most essential feature is their involvement in controlling morphogenetic processes, whose organization differs greatly in insects and vertebrates. Divergence in this direction, as compared to the reorganization of metabolic pathways, developed much faster and included the later stages of evolution. Thus it is not surprising that the corresponding regulatory systems also evolved relatively fast. This process reconstructed steroid hormones and the neuropeptides governing their synthesis; moreover, it determined the formation of entirely new inductors of morphogenetic transformations. In vertebrates, such new formations are evidently represented by thyroid hormones participating in the regulation of metamorphosis and other morphogenetic processes. In insects thyreoids and thyreotropic neuropeptides have not yet been detected. At the same time, insect postembryonic development is controlled by isoprenoid JH which have no analogues in the vertebrate endocrine system.

The divergent evolution of hormonal mechanisms regulating morphogenesis has certainly left its mark on the performance of other endocrine system functions as well, including the protective function. Nevertheless, even such polar groups of the animal kingdom as insects and vertebrates have retained a distinct parallelism of the main stress response elements associated with the secretion of biogenic amines and steroids. These elements must evidently be of a very ancient origin. Once emerged, they have evolved within rather narrow bounds, and in some cases they have remained almost unchanged. Hence it follows that the involvement of biogenic amines and some steroids in the regulation of the general adaptation syndrome must be a common rule for different groups of *Protostomia* and *Deuterostomia*.

VII. CONCLUSION

The up-to-date state of knowledge does not allow detailed analysis of the taxonomic, ontogenetic, and etiological diversity of insect hormonal reactions to stressor's action. Nevertheless, we shall try, even if in general line, to reconstruct the role of endocrine system in stress response and to present the way in which changes in the hormone titer lead to the survival of insects in extreme situations.

A stressor signal received by exteroreceptors is transmitted via sensory nerves into the brain whose neurones, having processed the information and assessed the stimulation level as extreme, transmit a message along cardiac nerves for an urgent release of neurohormones stored in corpora cardiaca.

The first reaction according to the speed of metabolic response is connected with the secretion of biogenic amines. Under their influence, within a few minutes the reserve energy substrates are mobilized in the fat body and in the nervous system. As a result, the readiness of the motor apparatus is achieved, an important condition for behavioral adaptation. If the switching of defensive forms of behavior does not remove the source of irritation, the organism is involved in a deeper state of stress, with adaptive and pathological shifts inherent in it. At this period biogenic amines often begin to play a negative role, causing a superexcitation and a subsequent paralysis of the nervous system.

At the next phase of the stress process a more complex problem of physiological adaptation is being solved. Its first part, i.e., providing energy, is partly realized by AKH and hypertrehalosemic peptides, in concert with biogenic amines or just following their action. It is possible that at a certain stage in stress development this group of hormones is joined by steroids which are related to glucocorticoids of vertebrates. The second part, i.e., stimulation of macromolecular synthesis and associated reparative process, is within the competence of ecdysteroids whose synthesis, in turn, is regulated by PTTH. In the same context of protein synthesis stimulation the possible protective function of testosterone-like steroids should be analyzed.

It is supposed that survival under extreme conditions is most closely associated with the maintenance of an intermediate ecdysteroid titer which surpasses the basal level characteristic of intact individuals during the intermolt period, but does not attain morphogenetically significant values. Thus, bringing ecdysteroid secretion into line with the needs of a damaged organism is a task more complex than in the case of other hormones. The simplest and most reliable way of solving this problem would be the existence of two forms of PTTH, one of which would induce a high morphogenetic activity of prothoracic glands and the other a moderate protective activity. With respect to the regulation of *Lepidoptera* metamorphosis, the existence of two such PTTH forms has been postulated by the hypothesis "two hormones—two functions",[21] but there is no proof of its applicability to the situation considered here.

Besides PTTH, ecdysteroid secretion can be modified by JH. In some cases JH stimulates ecdysteroids secretion and thereby promotes the activation of the effector systems subject to its control; in other cases JH creates prerequisites for a temporary delay of development. The field of direct metabolic JH effects connected with the regulation of protein and lipid metabolism also deserves further investigation within the framework of stress concept.

The involvement of hormones of an ecdysteroid or JH type in response to stress creates the danger of an untimely induction of morphogenetic process in target cells. Limiting the quantity of secreted hormones and shortening the period when target cells are sensitive to morphogenetic stimuli remove this danger. It seems probable that in some cases the consequences of inadequate cell response may be improved by some yet unknown reparative mechanisms whose functioning is promoted by delay in development.

One of the most intriguing peculiarities of the foregoing hypothetic scheme of insect endocrine response to the action of extraordinary stimuli is its surprising, though incomplete, similarity to that of vertebrates. Above all, it concerns the role of biogenic amines, steroid hormones, neuropeptides regulating steroidogenesis, and the metabolism of carbohydrates. Evidently, the mechanism of physiological adaptation organized as stress response had been formed, in general outline, at the earlier stages of Metazoa evolution, before the divergence of ancestor forms of insects and vertebrates took place.

REFERENCES

1. **Sternburg, J.**, Autointoxication and some stress phenomena, *Annu. Rev. Entomol.*, 8, 19, 1963.
2. **Martignoni, M. E.**, Pathophysiology in the insect, *Annu. Rev. Entomol.*, 9, 179, 1964.
3. **Perry, A. S. and Agosin, M.**, The physiology of insecticide resistance by insects, in *The Physiology of Insecta*, Vol. 6, Rockstein, M., Ed., Academic Press, New York, 1973, 3.
4. **Selye, H.**, Syndrome produced by diverse nocuous agents, *Nature*, 138, 32, 1936.
5. **Selye, H.**, *The Physiology and Pathology to Stress*, Acta, Montreal, 1950.
6. **Sabesan, S. and Ramalingam, N.**, Effects of endosulfan on the medial neurosecretory cells of adult male *Odontopus varicornis* (Dist.) *(Pyrrhocoridae: Heteroptera)*, *Entomon*, 4, 223, 1979.

7. **Kozlowskaya, V. I.**, The effect of phosphororganic insecticides on the neuroendocrine system of caterpillars of the cabbage moth *Mamestra brassicae L. (Lepidoptera, Noctuidae), Entomol. Rev.*, 49, 730, 1980 (in Russian).

8. **Prasad, O. and Srivastava, V. K.**, Effect of insecticide (BHC) on brain neurosecretory cells of *Poekilocerus pictus (Acrididae), Z. Mikrosk. Anat. Forsch.*, 94, 250, 1980.

9. **Tripathi, A. K.**, DDT-induced neurosecretory activity in *Edoiporus longicollis, Environ. Pollut.*, A28(2), 77, 1982.

10. **Balog, A. Ya., Guryeva, A. N., and Agutina, L. A.**, Effect of antibiotic novoimanin on the activity of CNS neurosecretory cells in insects, in *Antibiotics, Their Biological Role and Significance for Medicine and National Economy*, Naukova Dumka, Kiev, 1967, 51 (in Russian).

11. **Nanda, D., Ghosal, M. S., and Naskar, S.**, The effects of hypothermia and subsequent rewarming on the neurosecretory neuron in the brain of *Periplaneta americana, Folia Biol. (PRL)*, 25, 213, 1977.

12. **Ivanović, J. P., Janković-Hladni, M. I., and Milanović, M. P.**, Possible role of neurosecretory cells type A in response of *Morimus funereus* larvae to the effect of temperature, *J. Therm. Biol.*, 1, 53, 1975.

13. **Ivanović, J., Janković-Hladni, M., Stanić, V., and Milanović, M.**, The role of the cerebral neurosecretory system of *Morimus funereus* larvae *(Insecta)* in thermal stress, *Bull. T. 72 Acad. Serbe Sci. Arts Classe Sci. Nat. Math. Sci. Nat.*, 20, 91, 1980.

14. **Glumac, S., Janković-Hladni, M., Ivanović, J., Stanić, V., and Nenadović, V.**, The effect of thermal stress on *Ostrinia nubilalis HBN (Lepidoptera, Pyralidae)*. I. The survival of diapausing larvae, *J. Therm. Biol.*, 4, 277, 1979.

15. **Chernysh, S. I., Kudryavtseva, N. M., and Simonenko, N. P.**, unpublished data, 1988.

16. **Suzuki, A., Nagasawa, H., Kataoka, H., Hori, Y., Isogai, A., Tamura, S., Guo, F., Zhong, X., Ishizaki, H., Fujishita, M., and Mizoguchi, A.**, Isolation and characterization of prothoracicotropic hormone from silkworm Bombyx mori, *Agric. Biol. Chem.*, 46, 1107, 1982.

17. **Nishiitsutsuji-Uwo, J. and Nishimura, H. S.**, Partial purification and properties of prothoracicotropic hormone (PTTH) from developing adult brains of the silkworm *Bombyx mori, Insect Biochem.*, 14, 127, 1982.

18. **Ishizaki, H. and Suzuki, A.**, Prothoracicotropic hormone of *Bombyx mori:* primary structure and cellular localization as revealed by immunohistochemistry, *J. Cell Biochem.*, Suppl. 10C, 55, 1986.

19. **Bollenbacher, W. E. and Granger, N. A.**, Endocrinology of the prothoracicotropic hormone, in *Comprehensive Insect Physiology, Biochemistry and Pharmacology*, Vol. 7, Kerkut, G. A. and Gilbert, L. I., Eds., Pergamon Press, Oxford, 1985, 109.

20. **Ishizaki, H., Hizoguchi, A., Fujishita, M., Suzuki, A., Horiya, I., Ooka, H., Kataoka, H., Isogai, A., Nagasawa, H., Tamura, S., and Suzuki, A.**, Species specificity of the insect prothoracicotropic hormone (PTTH): the presence of Bombyx- and samia- specific PTTH's in the brain of *Bombyx mori, Dev. Growth Differ.*, 25, 593, 1983.

21. **Gilbert, L. I., Bollenbacher, W. E., Agui, W., Granger, N. A., Sedlak, B. I., Gibbs, D., and Buis, C. M.**, The prothoracicotropes: source of the prothoracicotropic hormone, *Am. Zool.*, 21, 641, 1981.

22. **Ishizaki, H.**, Changes in titer of the brain hormone during development of the silkworm *Bombyx mori, Dev. Growth Differ.*, 1, 11, 1969.

23. **Wigglesworth, V. B.**, The physiology of the cuticle and of ecdyses in *Rhodnius prolixus (Triatomidae, Hemiptera)* with special reference to the function of the oenocytes and of the dermal glands, *Q. J. Microsc. Sci.*, 76, 296, 1933.

24. **Nijhout, H. F.**, Stretch-induced molting in *Oncopeltus fasciatus, J. Insect Physiol.*, 25, 277, 1979.

25. **Clarke, K. U. and Langley, P. A.**, Studies on the initiation of growth and moulting in *Locusta migratoria migratorioides*, R. IV. The relationship between the stamotogastric nervous system and neurosecretion, *J. Insect Physiol.*, 9, 423, 1963.

26. **Pipa, R. L.**, Neuroendocrine involvement in the delayed pupation of space-deprived *Galleria mellonella (Lepidoptera), J. Insect Physiol.*, 17, 2441, 1971.

27. **Agui, N., Granger, N. A., Gilbert, L. Y., and Bollenbacher, W. E.**, Cellular localization of the insect prothoracicotropic hormone: *in vitro* assay of a single neurosecretory cell, *Proc. Natl. Acad. Sci. U.S.A.*, 76, 5694, 1979.

28. **Krishnakumaran, A.**, Injury induced molting in *Galleria mellonella* larvae, *Biol. Bull.*, 142, 281, 1972.

29. **Berreur, P., Porcheron, P., Berreu-Bonnenfant, J., and Dray, F.**, External factors and ecdysone release in *Calliphora erythrocephala, Experientia*, 35, 1031, 1979.

30. **Steel, J. E.**, Control of metabolic processes, in *Comprehensive Insect Physiology, Biochemistry and Pharmacology*, Vol. 8, Kerkut, G. A. and Gilbert, L. I., Eds., Pergamon Press, Oxford, 1985, 99.

31. **Samaranayaka-Ramasami, M.**, Insecticide-induced release of neurosecretory hormones, in *Pesticide and Venom Neurotoxic. 15th Int. Congr. Entomology*, Washington, D.C., 1978, 83.

32. **Maddrell, S. H. P.**, The insect neuroendocrine system as a target for insecticides, in *Insect Neurobiology and Pesticide Action (Neurotox 79) Proc. Soc. Chem. Ind. Symp.*, New York, 1979, London, 1980, 329.

33. **Singh, G. J. P. and Orchard, I.,** Is insecticide-induced release of insect neurohormones a secondary effect of hyperactivity of the central nervous system? *Pestic. Biochem. Physiol.,* 17, 232, 1982.

34. **Singh, G. J. R.,** Hormone release in *Locusta migratoria* in relation to insecticide poisoning syndrome, in *Insect. Neurochem. Neurophysiol. Proc. Int. Conf.,* College Park, MD, August 1 to 3, 1983, New York, London, 1984, 475.

35. **Van der Horst, D. S., Houben, N. M. D., and Beenakkers, A. M. T.,** Dynamics of energy substrates in the haemolymph of *Locusta migratoria* during flight, *J. Insect Physiol.,* 26, 441, 1980.

36. **Orchard, I., Loughton, B. G., and Webb, R. A.,** Octopamine and short-term hyperlipaemia in the locust, *Gen. Comp. Endocrinol.,* 45, 175, 1981.

37. **Kozanek, M., Slovak, M., Jurani, M., and Somogyiova, E.,** Influence of some exogenic factors on octopamine concentration and colouration of caterpillars of *Mamestra brassicae L. (Lepidoptera, Noctuidae), Biologia (Bratislava),* 40, 539, 1985.

38. **Colhoun, E. H.,** Some physiological and pharmacological effects of chlorinated hydrocarbon and organophosphorus poisoning, *Proc. N. Central Branch Entomol. Soc. Am.,* 14, 35, 1959.

39. **Hodgson, E. S. and Geldiay, S.,** Experimentally induced release of neurosecretory materials from roach corpora cardiaca, *Biol. Bull.,* 117, 275, 1959.

40. **Pence, R. J., Viray, M., Ebeling, W., et al.,** Honey-bee abdomen assays of hemolymph from stressed and externally poisoned American cockroaches, *Pestic. Biochem. Physiol.,* 5, 90, 1975; p. 539, 1985.

41. **Kozanek, M., Jurani, M., and Somogyiova, E.,** Influence of social stress on monoamine concentration in the central nervous system of the cockroach *Nauphoeta cinerea (Blattoidea), Acta Entomol. Bohemoslov.,* 83, 171, 1986.

42. **Fuzeau-Braesch, S.,** Contribution à l'étude in vivo du role neurophysiologique d'amines biogenes à l'aide d'un test de motilité chez un insecte subsocial *Locusta migratoria, C. R. Soc. Biol.,* 173, 558, 1979.

43. **Gole, J. W. D. and Downer, R. G. H.,** Elevation of adenosine 3'5'-monophosphate by octopamine in fat body of the American cockroach, *Periplaneta americana L., Comp. Biochem. Physiol. C,* 64, 223, 1979.

44. **Matthews, J. R. and Downer, R. G. H.,** Origin of trehalose in stress-induced hyperglycaemia in the American cockroach, *Periplaneta americana, Can. J. Zool.,* 52, 1005, 1974.

45. **Downer, R. G. H.,** Induction of hypertrehalosemia by excitation in *Periplaneta americana, J. Insect Physiol.,* 25, 59, 1979.

46. **Maddrell, S. H. P. and Reynolds, S. E.,** Release of hormones in insects after poisoning with insecticides, *Nature,* 236, 404, 1972.

47. **Jabbar, A. and Strang, R. H. C.,** Identity of a compound occurring in the blood of stressed insects, *Pestic. Sci.,* 16, 532, 1985.

48. **Hoffmann, J. A., Lagueux, M., Hetru, C., Kappler, C., Goltzene, F., Lanot, R., and Thiebold, J.,** Role of ecdysteroids in reproduction of insects: a critical analysis, in 4th *Int. Symp. Invertebrate* Reproduction, Lille, France, September 16, 1986.

49. **Norris, D. M., Rao, K. D. P., and Chu, H. M.,** Role of steroids in aging, in *Insect Aging,* Collatz, K., and Sochal, R. S., Eds., Springer-Verlag, Berlin, 1986, 182.

50. **Chernysh, S. I.,** Changes in ecdysone secretion in *Calliphora vicina* and *Barathra brassicae* under the influence of damage, *Biol. Sci.,* 5, 51, 1980 (in Russian).

51. **Karlson, P. and Shaaya, E.,** Der Ecdysontiter wahrend der Insectentwicklung. I. Eine Methode zur Bestimmung des Ecdysongehalts, *J. Insect Physiol.,* 10, 797, 1964.

52. **Chernysh, S. I., Lukhtanov, V. A., and Simonenko, N. P.,** unpublished data, 1985.

53. **Chernysh, S. I.,** *The Role of Ecdysones in the Physiological Adaptation to Damage in Insects,* Abstract of a thesis, Leningrad University Press, Leningrad, 1977 (in Russian).

54. **Girfanova, F. K.,** Changes in the resistance of larvae of *Calliphora vicina (Diptera, Calliphoridae)* to damaging factors in connection with diapause formation and cessation, *Zool. J.,* 63, 405, 1984.

55. **Chernysh, S. I., Girfanova, F. K., and Tyshchenko, V. P.,** The state of nonspecific resistance during insect diapause, *Proc. All Union Entomol. Soc.,* 68, 92, 1986 (in Russian).

56. **Kato, G. and Sakate, S.,** Early termination of summer diapause by mechanical shaking in pupae of Antheraea jamamai (Lepidoptera: Saturniidae), *Appl. Entomol. Zool.,* 18, 441, 1983.

57. **Lees, A. D.,** *The Physiology of Diapause in Arthropods,* Cambridge University Press, London, 1955.

58. **Beck, S. D.,** *Insect Photoperiodism,* 2nd ed., Academic Press, New York, 1980.

59. **Maslennikova, V. A. and Chernysh, S. I.,** The effect of ecdysterone on diapause determination in *Pteromalus puparum, Rep. U.S.S.R. Acad. Sci.,* 213, 480, 1973 (in Russian).

60. **Shappirio, D. G.,** Oxidative enzymes and the injury metabolism of diapausing cecropia silkworm, *Ann. N.Y. Acad. Sci.,* Art. 3(89), 537, 1960.

61. **Harvey, W. R. and Williams, C. M.,** The injury metabolism of the Cecropia silkworm. I. Biological amplification of the effects of localized injury, *J. Insect Physiol.,* 7, 81, 1961.

62. **Berry, S. J., Krishnakumaran, A., Oberlander, H., and Schneiderman, H. A.,** Effects of hormones and injury on RNA synthesis in saturniid moths, *J. Insect Physiol.,* 13, 1511, 1967.

63. **Schneiderman, H. A. and Williams, C. M.,** Physiology of insect diapause. VII. The respiratory metabolism of the Cecropia silkworm during diapause and development, *Biol. Bull.,* 105, 320, 1953.

64. **O'Farrell, A. F. and Stocks, A.,** Regeneration and the molting cycle in *Blatella germanica* L. I. Single regeneration initiated during the first instar, *Aust. J. Biol. Sci.,* 6, 485, 1953.

65. **Pohley, H. J.,** Experimentelle Beitrage zur Lenkung der Organentwicklung, des Hautungsrhytmus und der Metamorphose bei der Schabe *Periplaneta americana L., Wilhelm Roux Arch. Entwicklungsmech. Org.,* 151, 323, 1959.

66. **Roberts, B., Wentworth, S., and Kotzman, M.,** The levels of ecdysteroids in uninjured and leg-autotomized nymphs of *Blattela germanica* (L.), *J. Insect Physiol.,* 29, 679, 1983.

67. **Malevill, A. and de Reggi, M.,** Influence of leg regeneration on ecdysteroid titres in *Acheta* larvae, *J. Insect Physiol.,* 27, 35, 1981.

68. **Robertson, J. L. and Dell, T. R.,** Delayed metamorphosis in *Galleria mellonella:* interaction of chilling and space deprivation, *J. Insect Physiol.,* 27, 689, 1981.

69. **Shimizu, I., Adachi, S., and Kato, M.,** Acceleration of the time of moulting and larval ecdysis by electroshocks in the silkworm, *Bombyx mori, J. Sericult. Sci. Jpn.,* 47, 226, 1978.

70. **Chernysh, S. I.,** Response of the neuroendocrine system to damage action, in *Hormonal Regulation of Insect Development,* Giljarov, M. S., Tobias, V. J., and Burov, V. N., Eds., Nauka, Leningrad, 1983, 118 (in Russian).

71. **Piechowska, M. J.,** Effect of ionizing radiation on the endocrine system in insects, *Bull. Acad. Polon. Ser. Sci. Biol.,* 13, 139, 1965.

72. **Sivasubramanian, P., Ducoff, H. S., and Fraenkel, G.,** Effect of X-irradiation on the formation of the puparium in the flesh fly, Sarcophaga bullata, *J. Insect Phyisol.,* 20, 1303, 1974.

73. **Coulon, M. and Feyereisen, R.,** Effets de l'evolution du titre des ecdysteroides de l'hemolymphe et le metabolisme de l'ecdysone, *C. R. Acad. Sci. Ser. D,* 284, 1689, 1977.

74. **Madhavan, K. and Schneiderman, H. A.,** Hormonal control of imaginal disk regeneration in *Galleria mellonella (Lepidoptera), Biol. Bull.,* 137, 321, 1969.

75. **Pohley, H. J.,** Interactions between the endocrine system and the developing tissue in *Ephestia kuhniella, Behav. Sci.,* 15, 46, 1970.

76. **Chernysh, S. I., Lukhtanov, V. A., and Simonenko, N. P.,** Adaptation to damage in the silkworm *Bombyx mori (Lepidoptera, Bombycidae).* II. The effect of ecdysterone and other adaptiogenes on the resistance of caterpillars to entobacterin, *Entomol. Rev.,* 62, 665, 1983 (in Russian).

77. **Ito, Y., Shibazaki, A., and Iwahashi, O.,** Effect of phytoecdysones on the length of the fifth instar and the quality of cocoons in the silkworm, *Bombyx mori, Annot. Zool Jpn.,* 43, 175, 1970.

78. **Kobayashi, M. and Kawase, S.,** Multiplication of nuclear polyhedrosis virus in developing isolated pupae abdomens of the silkworm *Bombyx mori (Lepidoptera, Bombycidae), Appl. Entomol. Zool.,* 16, 307, 1981.

79. **Keeley, L. L. and Vinson, S. B.,** Beta-ecdysone effects on the development of nucleopolyhedrosis in *Heliothis* spp., *J. Invertebr. Pathol.,* 26, 121, 1975.

80. **Wei-Shan, C. and Horn-Sheng, L.,** Growth regulation and silk production in *Bombyx mori* L. from phytogenous ecdysteroids, in *Progress in Ecdysone Research,* Hoffmann, J. A., Ed., Elsevier/North-Holland, Amsterdam, 1986, 281.

81. **Chernysh, S. I., Lukhtanov, V. A., and Simonenko, N. P.,** Adaptation to damage in the silkworm *Bombyx mori L (Lepidoptera, Bombycidae).* III. Adaptiogenes and caterpillar's resistance to stressor activation of latent virus infection, *Entomol. Rev.,* 64, 267, 1985 (in Russian).

82. **Aruga, H.,** Induction of virus infections, in *Insect Pathology. An Advanced Treatise,* Vol. 1, Steinhaus, E. A., Ed., Academic Press, New York, 1963, 499.

83. **Karlson, P.,** Ecdysone in retrospect and prospect, in *Progress in Ecdysone Research,* Hoffmann, J. A., Ed., Elsevier/North-Holland, Amsterdam, 1980, 1.

84. **Terriere, L. C. and Yu, S. J.,** The induction of detoxifying enzymes in insects, *J. Agric. Food Chem.,* 22, 363, 1974.

85. **Terriere, L. C. and Yu, S. J.,** Interaction between microsomal enzymes of the house fly and the moulting hormones and some of their analogs, *Insect Biochem.,* 6, 109, 1976.

86. **Kobayashi, M. and Kimura, S.,** Action of ecdysone on the conversion of C^{14}-glucose in dauer pupa of the silkworm *B. mori, J. Insect Physiol.,* 13, 545, 1967.

87. **Day, M. F. and Oster, I. I.,** Physical injuries, in *Insect Pathology, An Advanced Treatise,* Vol. 1, Steinhaus, E. A., Ed., Academic Press, New York, 1963, 29.

88. **Gilbert, L. I., Bollenbacher, W. E., Goodman, W., Smith, S. L., Agui, N., Grenger, N., and Sedlak, B. J.,** Hormones controlling insect metamorphosis, *Rec. Prog. Horm. Res.,* 36, 401, 1980.

89. **Williams, C. M.,** Juvenile hormone. I. Endocrine activity of the corpora allata of the adult cecropia silkworm, *Biol. Bull.,* 116, 323, 1959.

90. **Krishnakumaran, A. and Schneiderman, H. A.,** Prothoracotrophic activity of compounds that mimic juvenile hormone, *J. Insect Physiol.,* 11, 1517, 1965.

91. **Nijhout, H. F. and Williams, C. M.,** Control of moulting and metamorphosis in the tobacco hornworm, *Manduca sexta (L.):* cessation of juvenile hormone secretion as a trigger for pupation, *J. Exp. Biol.,* 61, 493, 1974.

92. **Rountree, D. B. and Bollenbacher, W. E.,** The release of the prothoracicotropic hormone in the tobacco hornworm, *Manduca sexta,* is controlled intrinsically by juvenile hormone, *J. Exp. Biol.,* 120, 41, 1986.

93. **Hammock, B. D.,** Regulation of juvenile hormone titer: degradation, in *Comprehensive Insect Physiology, Biochemistry and Pharmacology,* Vol. 7, Kerkut, G. A. and Gilbert, L. I., Eds., Pergamon Press, Oxford, 1985, 431.

94. **Lusher, M. and Engelmann, F.,** Histologishe und experimentelle Untersuchungen uber die Auslosung der Metamorphose bei *Leucophaea maderae, Rev. Suisse Zool.,* 62, 649, 1960.

95. **Pohley, H. J.,** Untersuchungen uber die veranderung der Metamorphose rate durch Antennenamputation bei *Periplaneta americana, Wilhelm Roux Arch. Entwicklungsmech. Org.,* 153, 492, 1962.

96. **Retnakaran, A. and Bird, F. T.,** Apparent hormone imbalance syndrome caused by an insect virus, *J. Invertebr. Pathol.,* 20, 358, 1972.

97. **Subrahmanyam, B. and Ramakrishnam, N.,** The alteration of juvenile hormone titre in *Spodoptera litura (F.)* due to a baculovirus infection, *Experientia,* 36, 471, 1980.

98. **Bogus, M. and Cymborowski, B.,** Chilled *Galleria mellonella* larvae: mechanism of supernumerary moulting, *Physiol. Entomol.,* 6, 343, 1981.

99. **Raushenbach, I. J., Lukashina, N. S., and Korochkin, L. I.,** Genetic-endocrine regulation of the development of *Drosophila* under extreme environmental conditions. Communication. III. The effect of high culture density on the viability hormonal status and JH-esterase activity in *Drosophila virilis, Genetics,* 19, 1439, 1983 (in Russian).

100. **Raushenbach, I. J. and Lukashina, N. S.,** Genetic-endocrine regulation of the development of *Drosophila* under extreme environmental conditions. Communication. IV. Viability, hormonal status and activity of JH-esterase in *Drosophila virilis* during larval development on a poor diet, *Genetics,* 19, 1995, 1983 (in Russian).

101. **Terriere, L. C. and Yu, S. J.,** Insect juvenile hormones: induction of detoxifying enzymes in the housefly and detoxication by housefly enzymes, *Pestic. Biochem. Physiol.,* 3, 96, 1973.

102. **Boucias, D. G. and Nordin, G.,** Methoprene-nucleopolyhedrosis virus interactions in *Hyphantria cunea (Drury), J. Kang. Entomol. Soc.,* 53, 55, 1980.

103. **Chernysh, S. I. and Simonenko, N. P.,** The relationship between the resistance to injuries, level of metabolism and stability of the diapause in larvae of *Calliphora vicina (Diptera, Calliphoridae), Zool. J.,* 67, 1186, 1988 (in Russian).

104. **Burov, V. N., Kozhanova, N. I., and Reutskaya, O. E.,** The effect of hormone analogues on metamorphosis, reproduction, development and seasonal cycles, in *Hormonal Regulation of Insect Development,* Nauka, Leningrad, 1983, 140 (in Russian).

105. **Chernysh, S. I. and Kratina, N. D.,** Reversibility of hormone-induced damage of the imaginal development programme in *Musca domestica, J. Evol. Biochem. Physiol.,* 23, 684, 1987 (in Russian).

106. **Scharrer, E. and Scharrer, B.,** *Neuroendocrinology,* Columbia University Press, New York, 1963.

107. **Meola, S.,** The structural elements of insect neuroendocrine systems, *Southwest. Entomol.,* 8, 9, 1983.

108. **Girardie, A., Girardie, J., Lavenseau, L., Proux, J., Remy, C., and Vieillemaringe, J.,** Insect neurosecretion: similarities and differences with vertebrate neurosecretion, in *Neurosecretion and the Biology of Neuropeptides (Proc. 9th Int. Symp. on Neurosecretion),* Japanese Scientific Societies Press, Tokyo, 1985, 392.

109. **Scharrer, B.,** Insects as models in neuroendocrine research, *Annu. Rev. Entomol.,* 32, 1, 1987.

110. **Duve, H. and Thorpe, A.,** Comparative aspects of insect-vertebrate neurohormones, in *Insect Neurochemistry and Neurophysiology Proc. Int. Conf., College Park,* MD, August 1 to 3, 1983, New York, London, 1984, 171.

111. **Kramer, K. J.,** Vertebrate hormones in insects, in *Comprehensive Insect Physiology, Biochemistry and Pharmacology,* Vol. 7, Kerkut, G. A. and Gilbert, L. I., Eds., Pergamon Press, Oxford, 1985, 511.

112. **Maier, V., Grube, D., Pfeifle, B., Steiner, R., Greischel, A., and Pfeiffer, E. F.,** ACTH, gastrin, glucagon, and insulin immunoreactivities in the honey-bee, *Acta Endocrinol.,* 99, 36, 1982.

113. **Veenstra, J. A., Romberg-Privee, H. M., Schooneveld, H., and Polak, J. M.,** Immunocytochemical localization of peptidergic neurons and neurosecretory cells in the neuroendocrine system of the Colorado potato beetle with antisera to vertebrate regulatory peptides, *Histochemistry,* 82, 9, 1985.

114. **Hansen, B. L., Hansen, G. N., and Scharrer, B.,** Immunocytochemical demonstration of material resembling vertebrate ACTH and MSH in the corpus cardiacum-corpus allatum complex of the insect Leucophaea maderae, in *Handbook of Comparative Opioid and Related Neuropeptide Mechanism,* Stefano, G. B., Ed., CRC Press, Boca Raton, FL, 1986, 213.

115. **Verhaert, P. and de Loof, A.,** Substances resembling peptides of the vertebrate gonadotropin system occur in the central nervous system of *Periplaneta americana* L. Immunocytological and some biological evidence, *Insect Biochem.,* 16(Spec. Issue), 191, 1986.

116. **Nikitina, S. M.,** *Steroid Hormones in Invertebrates,* Leningrad University Press, Leningrad, 1982 (in Russian).

117. **de Clerck, D., Euchaute, W., Leuven, I., Diederik, H., and de Loof, A.,** Identification of testosterone and progesterone in hemolymph of larvae of the fleshfly *Sarcophaga bullata, Gen. Comp. Endocrinol.,* 52, 368, 1983.

118. **de Clerck, D., Diederic, H., and de Loof, A.,** Identification by capillary gaschromatography-mass spectrometry of eleven nonecdysteroid steroids in the haemolymph of larvae of *Sarcophaga bullata, Insect Biochem.,* 14, 199, 1984.

119. **El-Salhy, M., Falkmer, S., Kramer, K. J., and Speirs, R. D.,** Immunocytochemical evidence for the occurrence of insulin in the frontal ganglion of a lepidopteran insect the tobacco hornworm moth, *Manduca sexta* L., *Gen. Comp. Endocrinol.,* 54, 85, 1984.

120. **de Loof, A.,** New concepts in endocrine control of vitellogenesis and in functioning of the ovary in insects, in *Exogenous and Endogenous Influences on Metabolic and Neural Control,* Addink, A. D. E. and Spronk, N., Eds., Pergamon Press, New York, 1982, 165.

121. **de Loof, A.,** The impact of the discovery of vertebrate-type steroids and peptide hormone like substances in insects, *Entomol. Exp. Appl.,* 45, 105, 1987.

122. **Cannon, W. B.,** *The Wisdom of the Body,* Norton, New York, 1939.

123. **Axelrod, J. and Reisine, T. D.,** Stress hormones: their interaction and regulation, *Science,* 224, 452, 1984.

124. **Selye, H.,** *Hormones and Resistance,* Vol. 1 and 2, Springer-Verlag, Berlin, 1971.

125. **Selye, H.,** Adaptive steroids: retrospect and prospect, *Perspect. Biol. Med.,* 13, 343, 1970.

126. **Doane, W. W.,** Role of hormones in insect development, in *Developmental Systems: Insects,* Vol. 2, Counce, S. J. and Waddington, Eds., Academic Press, London, 1973, 291.

127. **Achrem, A. A., Levina, I. S., and Titov, Yu. A.,** *Ecdysones—Steroid Hormones in Insects,* Nauka, Minsk, 1973 (in Russian).

128. **Rosinski, G., Pilc, L., and Obuchowicz, L.,** Effect of hydrocortisone on the growth and development of larvae *Tenebrio molitor, J. Insect Physiol.,* 24, 97, 1978.

129. **Chernysh, S. I.,** Mammalian adrenocorticotropic hormone possesses the properties of insect prothoracicotropic hormone, *Rep. U.S.S.R. Acad. Sci.,* 271, 247, 1983 (in Russian).

130. **Ishizaki, H.,** A PTTH pilgrimage from *Luehdorfia* to *Bombyx, Zool. Sci.,* 3, 321, 1986.

131. **Arpagaus, M.,** Vertebrate insulin induces diapause termination in *Pieris brassicae* pupae, *Roux Arch. Dev. Biol.,* 196, 1527, 1987.

132. **Hochacka, P. W. and Somero, G. N.,** *Strategies of Biochemical Adaptation,* W.B. Saunders, Philadelphia, 1973.

133. **Kramer, K. J.,** Insulin-like and glucagon-like hormones in insects, in *Neurohorm. Techn. Insects,* New York, 1980, 116.

134. **Orchard, I. and Loughton, B. G.,** A hypolypemic factor from the corpus cardiacum of locusts, *Nature,* 286, 494, 1980.

135. **Lequellec, G., Gourdoux, L., Moreau, R., and Dutrieu, J.,** Regulation endocrinienne de la trehalosemie chez *Locusta migratoria:* comparaison des effets de deux extraits hormonaux endogenes avec ceux de l'insuline et du glucagon, *C. R. Acad. Sci. Ser. 3,* 294, 375, 1982.

136. **Nishiitsutsuji-Uwo, J. and Endo, Y.,** Immunohistochemical demonstration of brain-midgut endocrine system in the cockroach, in Insect Neurochemistry and Neurophysiology Proc. Int. Conf., College Park, MD, August 1 to 3, 1983, New York, 1984, 451.

Chapter 5

EFFECTS OF COLD STRESS ON ENDOCRINE SYSTEM OF *GALLERIA MELLONELLA* LARVAE

Bronisław Cymborowski

TABLE OF CONTENTS

I. INTRODUCTION

One of the thoroughly studied aspects of insect physiology is the effect of temperature on growth, development, and metamorphosis. In the optimal temperature environment most insect larvae develop at a fairly constant, species-specific rate, but adverse temperature conditions can affect both the number of molts and the length of consecutive instars. With increase of temperature, rate of growth and metabolism increase concomitantly.[1,2]

Essential to our understanding of temperature-mediated changes in insect physiology is knowledge of the impact that temperature has on the endocrine system. In the wax moth (*Galleria mellonella*) direct support for the existence of a temperature-sensitive endocrine mechanism comes from the finding that cold exposure (0°C) as short as a few minutes of the last instar larvae causes supernumerary moltings.[3,4] This is a unique phenomenon which has proved as a valuable tool in addressing many problems in physiology of *Galleria* with its atypical habitat in beehives, where changes in ambient temperature are considerably reduced. It is reasonable to assume that a temperature factor is affecting the developmental program of the endocrine system, but little is known of the pathways by which it exerts these effects.

It is the goal of this chapter to describe the developmental effects of cooling stress on the last instar *Galleria* larvae and elucidate the pathways by which these effects are exerted.

II. EFFECTS OF COLD STRESS ON THE NEUROENDOCRINE SYSTEM OF *GALLERIA MELLONELLA*

A. EFFECT OF COOLING STRESS ON SUPERNUMERARY MOLTS OF *GALLERIA MELLONELLA* LARVAE

When 1-d-old last instar *Galleria* larvae are exposed to 0°C the metamorphosis is inhibited and supernumerary instars are produced (Figure 1). The number of supernumerary instars depends on the cooling time used (Figure 2). Cooling 1-d-old larvae for a period shorter than 1 h stimulates additional larval instars in about 8%. Prolongation of cooling time to 2 h increases supernumerary molts up to 65%. The most effective cooling time is 3 h when incidence of supernumerary molts reaches 135%, which means that some of the cooled larvae molted twice. Further prolongation of cooling time does not increase incidence of supernumerary molts and a higher rate of mortality is observed.

It was also observed that some of the larvae subjected to low temperature pupated without spinning cocoons. The number of larvae incapable of cocoon spinning depends on the cooling time used: the longer the time the greater the number of larvae incapable of spinning. A total loss of spinning capacity occurs after 4 h of cooling (see Figure 2).

The third kind of developmental changes occurring in *Galleria* larvae due to cooling is the formation of individuals incapable of continuing normal development: these include dwarf larvae, abnormal pupae, and moths with abnormal wings. The number of these individuals also depends on the time of cooling applied (see Figure 2).

The susceptibility of *Galleria* larvae to cooling stress in the last instar decreases with age. The youngest larvae (within 2 d after the last ecdysis) form the greatest number of supernumerary instars. Older larvae either go into supernumerary instars in very small percentage, or not at all. After cooling, however, the number of individuals both incapable of spinning cocoons and of continuing normal development increases with age.[3]

In earlier instars (fourth, fifth, and sixth) there are no clear age-dependent changes in the sensitivity to cooling stress.[5]

Similar developmental changes can be obtained by juvenile hormone (JH) or juvenile hormone analogue (JHA) applications to the last instar of *Galleria* larvae (Table 1). When these compounds are applied to 1-d-old last instar larvae, supernumerary molts occur. When,

FIGURE 1. Supernumerary larval instar (left) obtained after 3 h of chilling (0°C) of 1-d-old last instar (right) of *G. mellonella*. (Magnification *circa* × 4.5.)

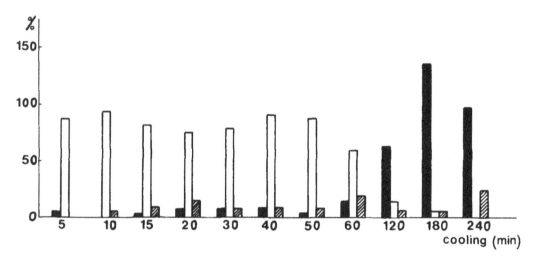

FIGURE 2. Changes in development of cooled 1-d-old last instar *Galleria* larvae. Black bars: percentage of supernumerary molts; open bars: percentage of larvae spinning cocoons; shaded bars: percentage of larvae incapable of continuing development. (Modified after Cymborowski and Bogus [1976].)

however, it is done to older larvae, they usually form a large number of individuals incapable of further development.[6] It may, therefore, be assumed that cooling exerts a juvenilizing effect by interfering with the hormonal system of an insect.

B. EFFECT OF COOLING ON JH LEVELS

Chilling stress during the first day of the last larval instar has a profound effect on the subsequent JH titers (Figure 3). During two consecutive days after chilling, JH levels were very high as measured by the "*Galleria* wax wound test" in comparison with unchilled controls. One day before ecdysis into supernumerary instar (on the third day after chilling) the JH titer is still elevated. In normally developing larvae at this time the JH level is undetectable (see Figure 3) due to an increase of specific juvenile hormone esterase (JHE) activity.[8,9]

Since allatectomized chilled *Galleria* larvae fail to produce supernumerary molts,[4] it

TABLE 1
Effect of Chilling Stress (0°C), Application of JH I or Hydroprene (ZR-512) on Supernumerary Molts of 1-d-old Last (Seventh) Instar *G. mellonella* Larvae

Treatment	Dose	Number of tested larvae	Percent of supernumerary molts
Controls	—	40	5.0
Chilling	3 h	137	70.0
	18 h	84	64.0
JH I	0.1 μg	22	9.1
	1.0	23	56.6
	10.0	42	97.6
	20.0	37	100.0
ZR-512	0.1 μg	32	6.3
	1.0	38	45.0
	10.0	114	99.9
	20.0	50	100.0

Note: Only molts to the eighth larval instars were counted.

After Muszynska-Pytel, personal communication.

seems obvious that a high JH titer is caused by activation of the *corpora allata* (CA). On the other hand, McCaleb and Kumaran[10] reported that JHE activity is low in cooled wax moth larvae. It is therefore possible that an increase in the JH titers observed in chilled larvae may be caused by both stimulation of the CA by the low temperature and a decrease in activity of JHE. In order to elucidate this, the activity of CA of both chilled and control (unchilled) larvae was measured.

C. *CORPORA ALLATA* ACTIVITY OF CHILLED *GALLERIA* LARVAE

It was shown that the CA activity, as measured by *in vitro* bioassay,[11] increases as a result of chilling stress application in comparison with unchilled control larvae (Figure 4). The highest synthetic activity of these glands occurs 24 h after cold application, whereas in control larvae the CA activity is at its lowest level at this time. The elevated CA activity of chilled larvae occurs even 48 h after chilling and is statistically different from those observed in controls (see Figure 4).

The questions which must now be addressed are how information about cooling stress is transmitted to an insect's endocrine system and what is the mechanism stimulating the CA activity after cooling stress application. A few reports directly implicate the role of the brain in these processes.

First of all it was shown that severing of the ventral nerve cord behind the subesophageal ganglion (SOG) on the third day of the last larval instar caused complete loss of the ability to produce supernumerary instars after subsequent chilling (Figure 5), although sensitivity to JH was preserved because about 50% of JHA-treated larvae whose ventral cord had been cut underwent supernumerary molts.[12] As can be seen from Figure 5 the severance of the nerve cord behind the last thoracic ganglion or between the fifth and sixth abdominal ganglia is considerably less effective in the prevention of supernumerary molts. This means that information about either physiological effect of chilling or perhaps about the actual temperature via thermoreceptors is transmitted to the brain via the nerve cord, and that the brain is responsible for integrating this information and then for turning the CA on. In further experiments it was shown that the brain allatotropic activity is involved in cold-induced supernumerary molts of *Galleria* larvae.

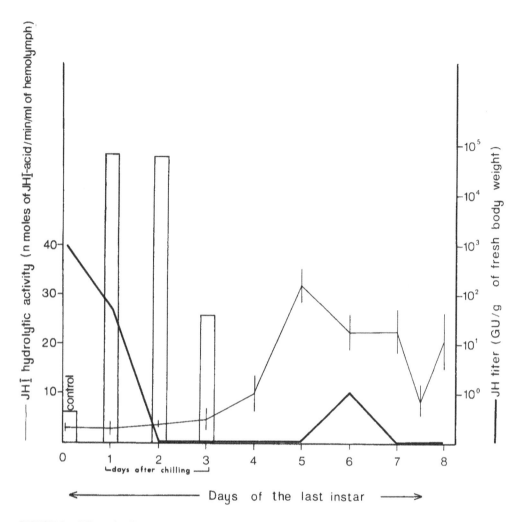

FIGURE 3. Effect of chilling stress on JH titers in the last instar *G. mellonella* larvae. Larvae were chilled (3 h at 0°C) on the first day after ecdysis to final instar. GU—*Galleria* unit. Each determination was performed on two pooled samples of *circa* 10 g of larvae (200 to 250 larvae). JH titers and JHE activity expressed in hydrolysis of JH I per milliliter of hemolymph per minute of control larvae are presented. Each point represents average value from three to seven independent measurements. Vertical bars indicate SEM. PP—formation of pharate pupae; P—pupation. (Based of the results of Cymborowski and Bogus [1981] and Bogus et al. [1987].)

D. EFFECT OF COOLING ON ALLATOTROPIC ACTIVITY OF THE BRAIN

The activity of brain was measured as the incidence of supernumerary molts by 1-d-old last instar *Galleria* larvae (recipient) implanted with brain dissected from either cooled or control larvae. Implantation of one brain taken from a donor on the second day after chilling results in about 50% of supernumerary molts (Table 2). The brain taken from an unchilled donor and implanted into a larva of the same age produces about 20% of supernumerary molts. In control experiments, when recipient larvae are implanted with the SOG taken from chilled donors or are only injured, the percentage of supernumerary molts is very low.

The activity of the brain taken from chilled larvae changes with respect to time after chilling (Figure 6). Implantation of brains dissected at the end of 3 h chilling period results in about 20% supernumerary molts, while the brains taken 18 h after cooling stimulate recipients in about 60% to produce supernumerary molts. The brain still has very high allatotropic activity 24 h after chilling, which then declines, and 48 h after chilling stress application there is no difference between chilled and unchilled larvae. On the second and third day after chilling the activity of brain decreased to the level observed just after chilling.[13]

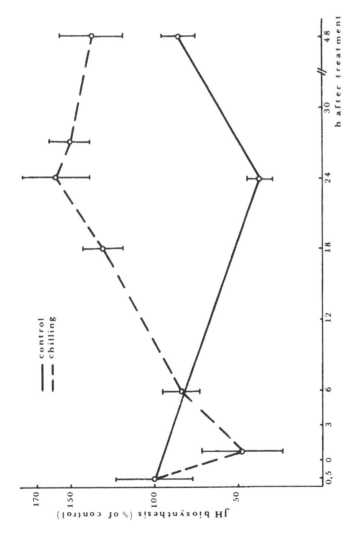

FIGURE 4.   The rate of *in vitro* JH synthesis (expressed in percent of controls) in *corpora allata-corpora cardiaca* glands of chilled (3 h at 0°C) and control 1-d-old last instar *G. mellonella* larvae. Glands were dissected 24 h after treatment. Each point represents mean value (± S.D.) of six replicate determinations. (After Muszynska-Pytel, personal communication.)

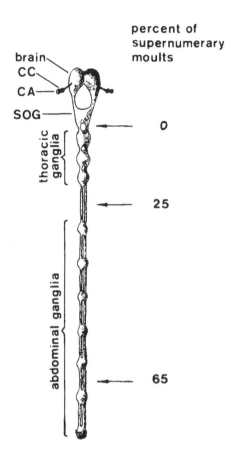

FIGURE 5.   Diagram of *G. mellonella* nervous system. Arrows indicate levels at which severings were performed. CC—corpus cardiacum; CA—corpus allatum; SOG—subesophageal ganglion.

TABLE 2
Effects of Various Treatments on Supernumerary Molts of the Last Instar *G. mellonella* Larvae[a]

Treatment	Number of replications	Fraction surviving	Percent of supernumerary molts
Chilled brain	4	38/40	52.6 = 8.1[b]
Unchilled brain	4	39/40	23.1 = 6.7[b]
Chilled SOG	3	26/26	0
Sham operation	3	25/25	8.0 = 5.4

[a] The brain and the subesophageal ganglion (SOG) were taken on the second day after chilling 1-d-old donor larvae and implanted into the 1-d-old host larvae. Results shown as a mean percent (SE).

[b] Data statistically significant by X^2 test at $p < 0.01$.

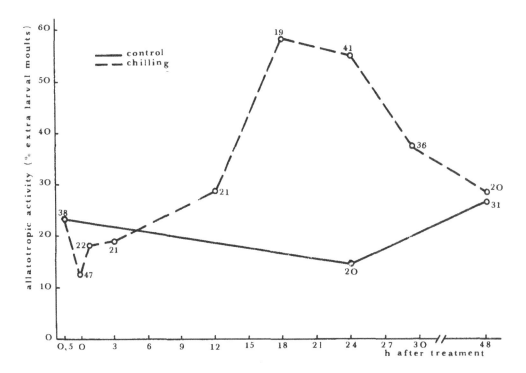

FIGURE 6.   Time-dependent allacotropic activity of the brain expressed as percentage of supernumerary molts. 1-d-old last instar larvae were chilled for 3 h, and the brains were dissected out at different times after termination of chilling and implanted into 1-d-old unchilled host larvae. The control larvae received the brains taken from unchilled donors. The numbers of brain recipients are indicated. (After Muszynska-Pytel, personal communication.)

The number of supernumerary molts produced by the host larvae progressively increases with respect to the number of implanted brains (Figure 7). When five brains were implanted into the host larva, some of them underwent supernumerary molts twice.

The same chilling stress (3 h at 0°C) has a different effect on the brain allatotropic activity depending on the age of cooled larvae. Brains of freshly molted last instar larvae have the highest allatotropic activity (Figure 8), but in this case chilling causes a very high mortality rate (about 80%). The activity then slightly declines but is still about two times higher as compared with the brains of the same age of unchilled larvae. In further experiments it was shown that an implanted brain must be present in the host larva for at least 2 d in order to induce supernumerary moltings, but it has no effect on the activity of the host brain.[13]

In further experiments it was shown that the brain has direct stimulatory effects on the CA biosynthetic activity. For this purpose an *in vitro* bioassay for measurement of allatotropic activity of *Galleria* brain was developed.[14]

In contrast to previously used methods of Ferenz and Diehl,[15] Gadot and Applebaum,[16] and Rembold et al.,[17] the allatotropic hormone (ATTH) was not extracted from the brain, but was obtained from the brain postincubation MEM medium. In this bioassay the activity of *corpora allata-corpora cardiaca* (CA-CC) complexes taken from the penultimate (sixth) instar *Galleria* larvae was measured.

It was found that the CA-CC complex is markedly activated by ATTH released by the brain taken from chilled 1-d-old last instar larvae as compared with unchilled controls (Figure 9). The rate of JH synthesis by CA-CC incubated in the presence of ATTH released from brains of the chilled last instar larvae was dose-dependent (Figure 10). The maximal activation (about 5.5-fold) was found for the glands incubated with ATTH of 2 brain equivalents (see Figure 10).

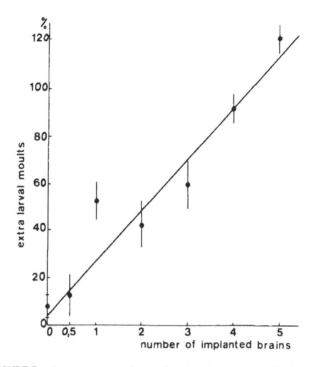

FIGURE 7.   Supernumerary molts as a function of the number of implanted brains. Brains were taken on the second day after 3 h chilling of 1-d-old last instar larvae and implanted into the 1-d-old unchilling host larvae. N = 25 − 40 in three replications. ±SE of mean value is indicated. (Reprinted with permission from Bogus, M. I. and Cymborowski, B., *J. Insect Physiol.*, 30, 557, 1984. Copyright 1984 Pergamon Press PLC.)

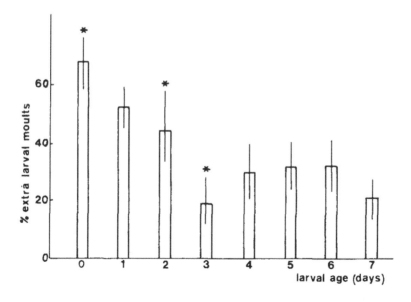

FIGURE 8.   Effect of age on the brain allatotropic activity of chilled last instar *Galleria* larvae. The brains were taken 1 d after chilling and implanted into 1-d-old unchilled host larvae. In all experiments each larva was supplied with one brain. N = 25 − 40 in three replications. ±SE of mean value is indicated. Asterisks indicate statistically significant data by X^2 test at $p = 0.05$ or better. (After Bogus and Cymborowski [1984].)

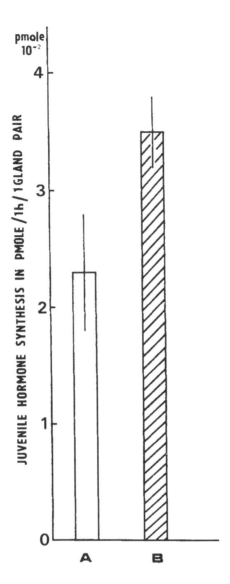

FIGURE 9.   The rate of *in vitro* JH synthesis by CA-CC glands of control (A) and chilled (B) 1-d-old *G. mellonella* larvae. Further explanations as for Figure 4. (After Muszynska-Pytel, personal communication.)

A detailed study performed by Muszynska-Pytel[18] has shown that the brain ATTH is released in *G. mellonella* from the median neurosecretory cells of the *pars intercerebralis*. These cells show the ability to elicit supernumerary larval molts after implantation into sensitive host larvae (Figure 11) and the ability to stimulate *in vitro* the JH synthesis by the CA-CC of *Galleria* larvae (Figure 12). At present further intensive studies are being carried out in order to isolate and characterize the *Galleria* ATTH.

On the basis of presented results a simple bioassay for screening the anti-JH compounds has been developed. It was found that the compounds known as potent inhibitors of JH biosynthesis, such as fluoromevalonate,[18] inhibit cold-induced supernumerary molts of *Galleria* larvae in a dose-dependent manner. This bioassay is also very convenient for investigation of the mechanism of action of different anti-JH compounds.[20]

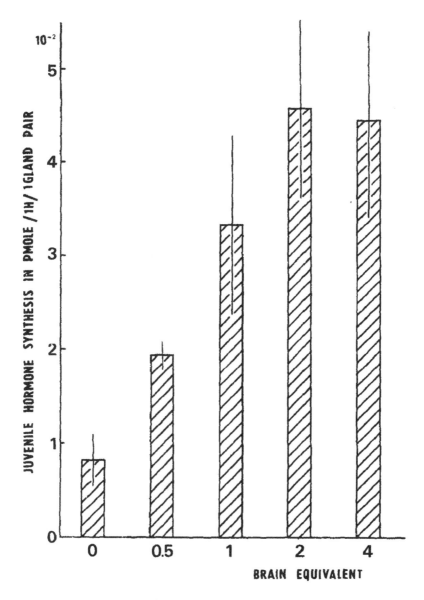

FIGURE 10. Dose response of CA-CC glands of *G. mellonella* larvae to allatotropic activity (ATTH) of the brain postincubation medium of 1-d-old last instar chilled larvae. Each bar represents mean value ($\pm$S.D.) of six to nine replicate determinations. (After Muszynska-Pytel, personal communication.)

E. EFFECT OF COOLING STRESS ON ECDYSTEROID LEVELS

In normally developing last instar *Galleria* larvae there are two peaks (about 132 and 168 to 174 h after final larval molt) in ecdysteroid titers.[21-23] In our rearing conditions (30°C) there is a small peak of ecdysteroids which occurs on day 6 of the last instar, whereas the main peak of ecdysteroid titers was observed on day 8, shortly before larval-pupal ecdysis (Figure 13). This peak was delayed by 1 d in comparison with that found by Sehnal et al. The pupation time was also delayed by 1 d.[24]

After cold treatment (3 h at 0°C) applied on the first day of the last instar an increase in hemolymph ecdysteroid titers was observed on day 4 (see Figure 13).

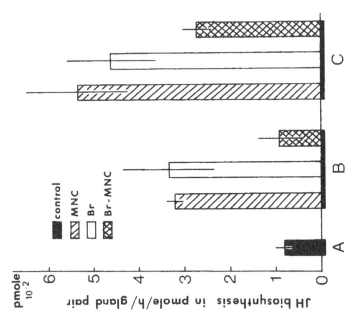

FIGURE 12.   The allatotropic activity of the median neurosecretory cells (MNC) of the brain of chilled 1-d-old last instar *G. mellonella* larvae, as tested by *in vitro* assay. The activity is expressed by the rate of JH synthesis by CA-CC of 3-d-old penultimate larvae *in vitro* stimulated by ATTH released from MNC or intact brain (Br) or else from MNC-deprived brain (Br-MNC). (A) Control CA-CC incubated with ATTH equivalent to: one cluster of MNC for one Br or else one Br-MNC per gland pair; CA-CC incubated with ATTH equivalent to two clusters of MNC or two Br or else two Br-MNC per gland pair. Each point represents the mean (SD) of three to six replicates. (From Muszynska-Pytel, M., *Experientia*, 43, 908, 1987. With permission.)

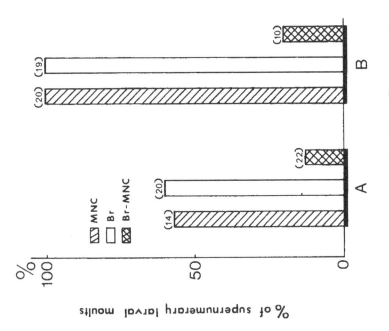

FIGURE 11.   The allatotropic activity of the median neurosecretory cells (MNC) of brain of chilled last instar *Galleria* larvae as tested by *in vivo* assay. The activity is expressed by the percentage of supernumerary larval molts of recipients upon implantation of: *A*-2 clusters of MNC (from two brains) or two intact brains (Br) or else two MNC-deprived brains (Br-MNC); *B*-4 clusters of MNC (from four brains) or four intact BR or else four Br-MNC. In parentheses are numbers of animals tested. (Fromr Muszynska-Pytel, M., *Experientia*, 43, 908, 1987. With permission.)

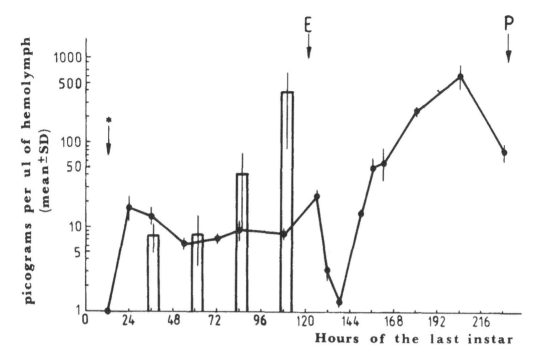

FIGURE 13. Changes in ecdysteroid titers (in pg of 20-OH-ecdysone equivalents per microliter of hemolymph) during normal development of the last instar *G. mellonella* larvae (continuous line) and after chilling stress application (3 h at 0°C) to 1-d-old larvae (bars). Ecdysteroid titers were determined by radioimmunoassay of 75% aqueous methanolic extracts of hemolymph. Each point was the mean of at least six separate determinations. Asterisk indicates period of chilling stress application. E—supernumerary larval ecdysis; P—pupation. (After Cymborowski et al., in preparation.)

When the samples for ecdysteroid measurements were taken every 6 h after cold treatment an oscillatory character in ecdysteroid titers could be detected before supernumerary molts, which usually occurred on day 5.[25]

It is obvious that chilling stress has a stimulatory effect on both allatotropic as well as prothoracicotropic activity of the brain. This is in accord with the suggestion of Pipa.[4]

Different results were obtained by Mala et al.[26] when the effect of cooling stress was investigated in larvae decapitated at 132 h of the last instar. Only about 20% of these larvae pupated within a month, whereas decapitated control (unchilled) larvae pupated in about 50% during the same period. According to authors these results suggest that cooling stress inhibits synthesis and/or release of ecdysteroids after decapitation. Since innervation of the prothoracic glands (PG) from the ventral cord[27] was intact in the decapitated larvae, the effect of chilling could have been exerted either directly on the glands or indirectly via the nerve cord.[28] It seems, therefore, that environmental stimuli affecting the neuroendocrine program which controls insect development exert their effect mainly via the brain, ventral nerve cord, and humoral factors.

III. ASSUMED MECHANISM OF CHILLING STRESS

On the basis of presented results the following regulating mechanism for cold-induced supernumerary moltings in *G. mellonella* larvae can be postulated (Figure 14). Information on the thermal state of the body is transmitted to the brain via the ventral nerve cord causing delay in its switching from a larval to pupal program. This leads to a release by the brain of an allatotropin (ATTH) which stimulates the CA to produce a high level of JH. Evidence

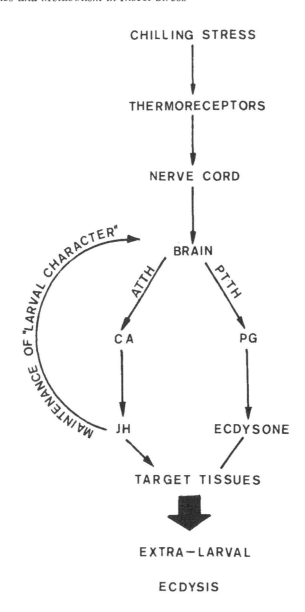

FIGURE 14. Suggested mechanism of cold-induced supernumerary molts of the last instar *G. mellonella* larvae. ATTH—allatotropic hormone; CA—corpora allata; PG—prothoracic glands; JH—juvenile hormone. Further explanations in the text.

for a humoral factor stimulating the activity of the CA comes from experiments in which implantation of the brain from chilled larvae into freshly molted last instar recipients resulted in a higher incidence of supernumerary molts than in control larvae receiving the brains of unchilled larvae[13] and from *in vitro* stimulation of JH synthesis by the brain.[18] Since the brain of chilled larvae maintains its larval character, it can release PTTH which stimulates PG to produce ecdysone in the presence of JH, and supernumerary larval ecdysis will be induced.

In some cases chilled larvae produce not only one, but two or even three supernumerary instars.[3] It seems probable that some of the last instar chilled larvae ecdyse to superlarvae having much higher JH titers than others, since JH levels on the third day (1 d before supernumerary molt) are still very high. A higher JH level might prevent "reprogramming" of the brain, so the larval cycle is repeated again in these animals. In addition, the presence of high levels of JH in chilled last instar larvae can prevent the PG from undergoing the switchover in their sensitivity to JH. As shown in *Spodoptera littoralis*[29] and *Manduca sexta*,[30] JH has an inhibitory effect shortly before pupation. As long as JH is present in the larvae, the PG cannot undergo a switchover, and their program remains larval. It is worth noting that high levels of JH in chilled larvae could prevent both the brain and the PG from changing their program from larval to pupal.

It has been shown that treatment of *M. sexta* larvae with hydroprene caused the CA to become inactivated, as measured by the subsequent loss of the ability to synthetize JH *in vitro*,[31] and a substantial decrease in JH titer.[32,33] The latter authors have suggested that the reduction in the endogenous JH level resulting from treatment of *Manduca* larvae with hydroprene is due to the effect of this compound on the feedback mechanism responsible for switching the CA off and on. In the case of *G. mellonella* the CA activity seems to be regulated by ATTH via a positive feedback mechanisms. This is supported by the fact that in larvae having the high JH level, the allatotropic activity of the brain is also very high. On the contrary, the low rate of JH synthesis is accompanied by low allatotropic activity of the brain.

ACKNOWLEDGMENT

This work was supported in part by scientific program RP-II-12 coordinated by the Jagiellonian University, Cracow.

REFERENCES

1. **Reynolds, S. E. and Nottingham, S. F.,** Effect of temperature on growth and efficiency of food utilization in fifth-instar caterpillars of the tobacco hornworm, *Manduca sexta, J. Insect Physiol.,* 31, 129, 1985.
2. **Roe, R. M., Clifford, C. W., and Woodring, J. P.,** The effect of temperature on energy distribution during the last larval stadium of the female house cricket, *Acheta domesticus, J. Insect Physiol.,* 31, 371, 1985.
3. **Cymborowski, B. and Bogus, M. I.,** Juvenilizing effect of cooling on *Galleria mellonella, J. Insect Physiol.,* 22, 669, 1976.
4. **Pipa, R. L.,** Supernumerary instars produced by chilled wax moth larvae: endocrine mechanisms, *J. Insect Physiol.,* 22, 1641, 1976.
5. **Bogus, M. I. and Cymborowski, B.,** Effect of cooling stress on growth and developmental rhythms in *Galleria mellonella, Bull. Acad. Pol. Sci.,* 25, 257, 1977.
6. **Sehnal, F. and Schneiderman, H. A.,** Action of the corpora allata and of juvenilizing substances on the larval-pupal transformation of *Galleria mellonella* L. (Lepidoptera), *Acta Entomol. Bohemoslov.,* 70, 289, 1973.
7. **de Wilde, J., Staal, G., de Kort, C. A. D., de Loof, A., and Baard, G.,** Juvenile hormone titers in the haemolymph as function of photoperiodic treatment in the adult Colorado beetle (*Leptinotarsa decemlineata,* Say.), *Proc. K. Ned. Akad. Wet.,* 71, 321, 1968.
8. **Sehnal, F. and Rembold, H.,** Brain stimulation of juvenile hormone production in insect larvae, *Experientia,* 41, 684, 1985.
9. **Bogus, M. I., Wisniewski, J. R., and Cymborowski, B.,** Effect of lighting conditions on endocrine events in *Galleria mellonella, J. Insect Physiol.,* 33, 355, 1987.
10. **McCaleb, D. C. and Kumaran, A. K.,** Control of juvenile hormone esterase activity in *Galleria mellonella* larvae, *J. Insect Physiol.,* 26, 171, 1980.

11. **Wisniewski, J. R., Muszynska-Pytel, M., Grzelak, K., and Kochman, M.,** Biosynthesis and degradation of juvenile hormone in corpora allata and imaginel discs of *Galleria mellonella, Insect Biochem.,* 17, 249, 1987.

12. **Bogus, M. I. and Cymborowski, B.,** Chilled *Galleria mellonella* larvae: mechanism of supernumerary moulting, *Physiol. Entomol.,* 6, 343, 1981.

13. **Bogus, M. I. and Cymborowski, B.,** Induction of supernumerary moults in *Galleria mellonella:* evidence for an allatotropic function of the brain, *J. Insect Physiol.,* 30, 557, 1984.

14. **Muszynska-Pytel, M., Szolajska, E., and Wisniewski, J. R.,** In vitro allatotropic activity of *Galleria mellonella* (Lepidoptera, Insecta) larval brain, in preparation.

15. **Ferenz, H. J. and Diehl, I.,** Stimulation of juvenile hormone biosynthesis in vitro by locust allatotropin, *Z. Naturforsch. Teil C,* 38, 856, 1983.

16. **Gadot, M. and Applebaum, S. W.,** Rapid in vitro activation of corpora allata by extracted locust brain allatotropic factor, *Arch. Biochem. Physiol.,* 2, 117, 1985.

17. **Rembold, H., Schlagintweit, B., and Ulrich, G. M.,** Activation of juvenile hormone synthesis in vitro by a corpus cardiacum factor from *Locusta migratoria, J. Insect Physiol.,* 32, 91, 1986.

18. **Muszynska-Pytel, M.,** Allatotropic activity of the median neurosecretory cells of *Galleria mellonella* (Lepidoptera) larval brain, *Experientia,* 43, 908, 1987.

19. **Quistad, G. B., Derf, D. C., Schooley, D. A., and Staal, G. B.,** Fluoromevalonate acts as inhibitor of juvenile hormone biosynthesis, *Nature,* 289, 176, 1981.

20. **Malczewska, M. and Cymborowski, B.,** Cold-induced supernumerary moults of *Galleria mellonella* larvae as bioassay for anti-juvenile hormone compounds, *Acta Physiol. Pol.,* 39, 112, 1988.

21. **Bollenbacher, W. E., Zvenko, H., Kumaran, A. K., and Gilbert, L. I.,** Changes in ecdysone content during postembryonic development of the wax moth, *Galleria mellonella:* the role of the ovary, *Gen. Comp. Endocrinol.,* 34, 169, 1978.

22. **Sehnal, F., Maroy, P., and Mala, J.,** Regulation and significance of ecdysteroid titre fluctuations in lepidopterous larvae and pupae, *J. Insect Physiol.,* 27, 535, 1981.

23. **Sehnal, F., Delbecque, J. P., Maroy, P., and Mala, J.,** Ecdysteroid titers during larval life and metamorphosis of *Galleria mellonella, Insect Biochem.,* 16, 157, 1986.

24. **Cymborowski, B., Smietanko, A., and Delbecque, J. P.,** Circadian modulation of ecdysteroid titer in *Galleria mellonella* larvae, *Comp. Biochem. Physiol.,* 94A, 431, 1989.

25. **Malczewska, M., Gelman, D. B., and Cymborowski, B.,** Effect of azadirachtin on development, juvenile hormone and ecdysteroid titers in chilled *Galleria mellonella* larvae, *J. Insect Physiol.,* 34, 725, 1988.

26. **Mala, J., Sehnal, F., Krishna Kumaran, A., and Granger, N. A.,** Effect of starvation, chilling and injury on endocrine gland function in *Galleria mellonella, Arch. Insect Biochem. Physiol.,* 4, 113, 1987.

27. **Singh, H. H. and Sehnal, F.,** Lack of specific neurons in the ventral nerve cord for the control of prothoracic glands, *Experientia,* 35, 1117, 1979.

28. **Mala, J. and Sehnal, F.,** Role of the nerve cord in the control of prothoracic glands in *Galleria mellonella, Experientia,* 34, 233, 1978.

29. **Cymborowski, B. and Stolarz, G.,** The role of juvenile hormone during larval-pupal transformation of *Spodoptera littoralis:* switch-over in the sensitivity of the prothoracic glands to juvenile hormone, *J. Insect Physiol.,* 25, 939, 1979.

30. **Safranek, L., Cymborowski, B., and Williams, C. M.,** Effects of juvenile hormone on ecdysone-dependent development in the tobacco hornworm, *Manduca sexta, Biol. Bull.,* 158, 248, 1980.

31. **Kramer, S. J. and Staal, G. B.,** In vitro studies on the mechanism of action of anti-juvenile hormone agents in larva of *Manduca sexta,* in *Juvenile Hormone Biochemistry,* Pratt, G. E. and Brooks, G. T., Eds., Elsevier/North-Holland, Amsterdam, 1981, 425.

32. **Edwards, J. P., Bergot, B., and Staal, G. B.,** Effects of three compounds with anti-juvenile hormone activity and a juvenile hormone analogue on endogenous juvenile hormone levels in the tobacco hornworm, *Manduca sexta* larvae, *J. Insect Physiol.,* 29, 83, 1983.

33. **Baker, F. C., Miller, C. A., Tsai, L. W., Jamieson, G. C., Cerf, D. C., and Schooley, D. A.,** The effects of juvenoides anti-juvenile hormone agents, and several intermediates of juvenile hormone biosynthesis on the in vivo juvenile hormone levels in *Manduca sexta* larvae, *Insect Biochem.,* 16, 741, 1986.

Chapter 6

CHANGES IN JUVENILE HORMONE AND ECDYSTEROID CONTENT DURING INSECT DEVELOPMENT UNDER HEAT STRESS

Inga Y. Rauschenbach

TABLE OF CONTENTS

I. INTRODUCTION

Unfavorable environmental conditions may affect insect development in different ways, i.e., induce a facultative diapause, delay growth, and/or prevent metamorphosis in insects developing without diapause.[1]

A. INSECT DEVELOPMENT UNDER HEAT STRESS

In 1952, Wigglesworth[2] demonstrated the effects of high temperature (above the optimum temperature for the given species) on larval development of *Rhodnius prolixus*. He stated that an exposure of fed last instar larvae to 35°C delayed the development of all the larvae and prevented molts into adults in 50% of them. However, resumed feeding of the latter and their subsequent transfer to the control temperature induced the molting activity and adult development. An analogous response to the effect of high temperature was also observed in other insect species. Mellanby[3] has reported that under the effect of high temperatures the metamorphosis of larval *Aedes aegypti* and *Tenebrio molitor* was delayed almost by a month. It was Church[4] who demonstrated that in larvae of *Cephus cinctus* an exposure to high temperature slowed down their development and increased the larval mortality. Vardell and Tilton[5] have shown that development of *Plodia interpunctella* larvae exposed to 34°C was delayed and 82 to 90% of them died.

We also investigated the effect of high temperature (32°C) on the development of *Drosophila virilis* and attempted to find out whether or not there exists a critical period during development susceptible for the induction of delayed metamorphosis under the above temperature conditions.

Synchronized *D. virilis* cultures were heat-exposed during the embryonic development, the first, second, and/or third larval instars, each of them separately. After completion of each stage of development, the experimental cultures were transferred to the control temperature (25°C).

It was found that the development of first and second larval instars of *D. virilis* at 32°C was not affected by this temperature in relation to the metamorphosis and the survival rate (Table 1). However, when third instar larvae developed at 32°C, the onset of metamorphosis was delayed by 12 to 14 h (Table 2) and some of the larvae died (Table 1). Hence, it was concluded that in *D. virilis* the developmental period critical for the stressful effect of a high temperature is the third larval instar.

Thus, on the basis of the data available, it is evident that under the stressful effect of high temperatures the molt is delayed or prevented in Hemimetabola (*R. prolixus*) and Holometabola (*A. aegypti, T. molitor, C. cinctus, P. interpunctella, D. virilis*).

B. HORMONAL CONTROL OF INSECT DEVELOPMENT UNDER HEAT STRESS

Wigglesworth[2,6] was the first who supposed that in insects exposed to the effect of

TABLE 1
Effect of Exposure to High Temperature during Different
Developmental Stages on the Survival of *D. virilis* Individuals

Stage	32°C		25°C	
	Number of individuals	Survival[a] (%)	Number of individuals	Survival[a] (%)
Embryogenesis	740	57 ± 1.8	892	56 ± 1.7
First larval instar	125	95 ± 1.9	125	97 ± 1.5
Second larval instar	122	96 ± 1.8	126	93 ± 1.2
Third larval instar	135	71 ± 3.9	137	92 ± 2.3

[a] Survival was expressed as the ratio of the number of emerged adults to the number of treated individuals.

unfavorable temperatures delayed metamorphosis is the consequence of a disturbance in hormonal balance. Mellanby,[3] who studied the effect of high temperature on larvae of *A. aegypti*, and Church,[4] on those of *C. cinctus*, suggested that delayed development might be caused by the absence of ecdysone.

Ten years later, O'Kasha[7-9] was able to explain the first link in the processes leading to delay or prevention of molting under the effect of a sublethal high temperature. In experiments on *R. prolixus*, he severed the larval brain before and after the critical period, i.e., before and after the secretion of prothoracicotropic hormone (PTTH). Brainless larvae were thereafter exposed to 36°C. Molt occurred only in those larvae whose brains were removed after the critical period. On the basis of these observations, O'Kasha[7-9] concluded that delay or prevention of molting at sublethal high temperature might be caused by the cessation of PTTH production and, as a consequence, by the suppression of ecdysone synthesis.

In a series of publications Ivanović et al.[10,11] and Janković-Hladni et al.[12] have demonstrated the effects of stressful temperatures on the neurosecretory system of two species of larval Cerambycidae (Coleoptera). In most studies, larvae of *Morimus funereus*, developing 3.5 to 4 years under natural conditions, were used. Larvae of different age and collected during different seasons were exposed to the following constant temperatures: −1, +8, and/or +23°C. The response of the larvae to the effect of stressful temperatures was dependent on the phase of larval development and the phase of the annual cycle. In relation to the biochemical and cytological parameters studied, the most sensitive to the effect of stressful temperatures were protocerebral A1 neurosecretory cells of the median group. It was suggested that these neurosecretory cells (NSC) synthesize the PTTH. Janković-Hladni et al.[12] have shown that under the effect of different factors, the protocerebral NSCs of the larval *Cerambyx cerdo* responded selectively.

Our results obtained in experiments with the fruit fly *D. virilis* have also shown a decrease in PTTH secretion under the effect of high temperature. The above was estimated first, according to the decrease in the neurosecretory activity of the NSC secreting the PTTH and their concomitant decrease in size,[13] and second, in relation to the results of a bioassay concerning the determination of PTTH content in the brain homogenates of larvae grown either at 32 or 25°C. The bioassay of PTTH content was based on differences in the rate of development of recipient larvae injected with brain homogenates prepared from heat-treated and/or control larvae. The brains used for homogenates were severed 12 h before pupariation of larvae took place. The results shown in Table 3 demonstrate the presence of PTTH in the brains of the controls (an accelerated metamorphosis in recipients) and a sharp decline in its level in brains of the heat-treated larvae, i.e., there are no differences in the rate of metamorphosis between recipients of the brain homogenate of heat-treated larvae and those of the Beadle medium.

TABLE 2
Effect of High Temperature on the Developmental Time of *D. virilis* Larvae Heat-Exposed during the Third Instar

Group of larvae	Number of larvae	Pupation time (hours after the second molt)								Larvae nonpupating at 32°C (%)	Dead larvae (%)
		72	76	80	84	88	92	96	100		
		% of larvae pupated for time period									
Controls	82	7	40	37	13	0	0	0	0	0	3
Heat-exposed (32°C)	82	0	0	0	7	23	21	11	9	20[a]	9

[a]   These larvae pupated when the cultures were transferred to 25°C.

TABLE 3
**The Effect of Injections of Brain Homogenates Prepared from Heat-Exposed (32°C)
and Control (25°C) _D. virilis_ Larvae on the Developmental Time of Recipients**

Groups of larvae	Number of larvae	% Larvae pupariated after					Dead larvae (%)
		24 h	32 h	40 h	48 h	56 h	
Control, injected with Beadle medium	43	0	14	26	33	7	20
Injected with the brain homogenate from heat-exposed larvae	41	0	14.7	21.9	34	14.7	14.7
Injected with the brain homogenate from control larvae	51	20	25	33	14	0	8

The authors who studied the causes of retardation and/or prevention of molt under heat stress agreed about the supposition that a drastic decrease in PTTH secretion evokes the suppression of ecdysone secretion.[2-4,6-9] The above was supported by experiments of Wigglesworth[6] and O'Kasha[7] concerning the removal and reimplantation of the prothoracal glands (PG) into heat-treated larvae of _R. prolixus_.

We also attempted to measure ecdysteroid titer in the heat-stressed _D. virilis_ larvae.

II. CHANGES IN ECDYSTEROID TITER UNDER HEAT STRESS

As already mentioned, all the authors who studied the effect of heat stress on insect development have shown that stressful temperature elicited different responses in insects: some individuals of each species tested, after a delay in development, underwent metamorphosis and survived, while the remainders unable to metamorphose died.

We attempted to determine ecdysteroid content in both types of larvae of _Drosophila virilis_ under heat stress. For this purpose, we measured the ecdysteroid content during the stadium of wandering larvae, shortly before the onset of metamorphosis. However, at this larval state it is impossible to distinguish between larvae able and those unable to metamorphose when heat-stressed. Therefore, lines of insects either able or unable to undergo metamorphosis at high stressful temperature were needed. From eight lines of _D. virilis_ of an independent origin, two lines contrasting in their ability to undergo metamorphosis at 32°C were used in experiments. Line 101 consisted of larvae which, after a delay in development of 12 to 14 h, completed their metamorphosis. The larvae of line 147 were unable to pupate at 32°C, but after their transfer to the control temperature (25°C) they all pupated, but never developed into adults and all died at the stage of pupa.

The results concerning determinations of PTTH content in larvae of lines 101 and 147 have demonstrated a decrease in PTTH secretion in both lines.[13] Hence, one may expect a reduction in ecdysteroid secretion, but this was not the case.

A. THE STATE OF PG IN _D. VIRILIS_ LARVAE ABLE AND/OR UNABLE TO METAMORPHOSE UNDER HEAT STRESS

The size of glands secreting insect hormones, their cells, and cell nuclei allows an indirect evaluation of the relative hormone content secreted under the effect of extrinsic factors. In a series of publications it was established that the size of corpus allatum (CA) and PG reflects their functional activities.[14-19] It must be mentioned that such comparative estimates are valid only in those cases when the compared insects are of an identical developmental stage.[20]

Figure 1 represents the sizes of the PGs and their nuclei in wandering larvae of lines

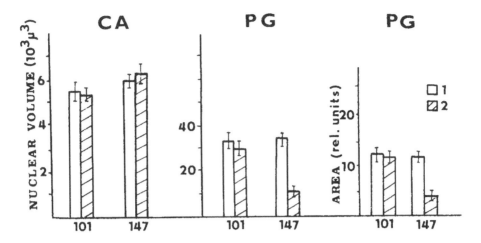

FIGURE 1.   Effect of high temperature on the nuclear volume CA and PG, and also on the PG surface area in *D. virilis* larvae of lines 101 and 147. 1—control (25°C), 2—heat-exposed (32°C).

101 and 147 at high (32°C) and control (25°C) temperatures. In larvae of line 147 exposed to 32°C the volume of nuclei of PG cells was significantly decreased when compared to the nuclei of the controls. The same holds true for the PG surface area. In larvae of line 101 exposed to the same temperature there were no significant differences between the heat-treated and control larvae. Thus, the results show that the PGs are active in larvae able to undergo metamorphosis when heat-stressed (line 101), which was also demonstrated by their capability to pupate at 32°C (the reason why ecdysone is present in the absence of PTTH in these larvae will be discussed later). In larvae of line 147, unable to metamorphose at 32°C, the PG seems to be inactive and consequently the ecdysone synthesis must be sharply decreased in comparison with the controls.

It must be recalled that there exists mutants of *D. melanogaster* in which the PG is hypertrophied and at the same time the synthesis of ecdysone is decreased.[21,22] In order to exclude or confirm such a possibility in the case of *D. virilis*, we measured the ecdysteroid level in the PGs of heat-treated and control larvae of line 147 using a biossay.

B. RELATIVE ECDYSONE CONTENT IN THE PG OF *D. VIRILIS* LARVAE UNABLE TO METAMORPHOSE UNDER HEAT STRESS

It is known that injection of ecdysone into larvae of *D. melanogaster* in the middle of the last instar evokes cuticle tanning (formation of the pupal epidermis in larvae) or precocious pupation.[23]

Our bioassay was performed as follows: ring gland homogenates of wandering heat-treated and control larvae of line 147 were injected into control larvae of the same line 24 h before pupariation. Two additional groups of larvae were injected either with ecdysterone or with Beadle medium. The larvae injected with Beadle medium served as a control for Beadle medium in which the ring gland homogenates were prepared and ecdysterone dissolved. After the injection all recipients as well as a group of intact larvae were maintained at 25°C. The results of the bioassay are shown in Table 4.

The injection of ring gland homogenates prepared from control larvae and of ecdysterone has evoked identical responses in the recipients, i.e., 10 h after the injection, larvae with pupal epidermis as well as some pupae appeared in both groups of recipients. In the course of 24 h following the treatment almost all the larvae pupated. Hence, it was concluded that PGs of larvae grown at 25°C synthesize ecdysone. On the other hand, injections of the ring gland homogenates prepared from heat-stressed larvae and the Beadle medium alone did not accelerate the rate of development of *D. virilis* larvae. The above results are probably

TABLE 4
Effects of Injections of Ecdysterone and Ring Gland Homogenates on the Cuticle and Pupation Time in Recipients

Group	Number of larvae	Time after injection (h)				Dead larvae (%)
		10	10	24	36	
		Larvae with pupal epidermis (%)	Pupae (%)			
I	54	35	26	20	6	13
II	46	46	17.5	17.5	9	10
III	43	0	0	13	78	9
IV	43	0	0	12	75	13
V	45	0	0	69	31	0

Designations: I—ecdysterone injection; II—injection of ring gland homogenates from control larvae; III—injection of ring gland homogenates from heat-exposed larvae; IV—injection of Beadle medium; V—intact controls.

TABLE 5
The Effect of an Exogenous Ecdysone Supply on the Developmental Time of *D. virilis* at 25°C

Group of larvae	Number of larvae	Pupation time (hours after the second molt)			
		48	72	96	120
		Larvae pupated during the time (%)			
Control	60	0	7	83	10
Ecdysone-treated	65	31	31	34	4

the consequence of extremely low content of ecdysone in the ring gland of heat-treated larvae.

It is interesting to note that the injection itself, i.e., pricking, retards the onset of metamorphosis: larvae injected with Beadle medium pupated later than the intact controls. This is in agreement with the results of O'Kasha[7-9] who observed delayed metamorphosis in *Rhodnius prolixus* larvae after doing damage to their cuticle.

Hence, it may be concluded that the ecdysone level in PGs of larvae unable to undergo metamorphosis under the heat stress is drastically reduced.

We attempted to compensate for this endogenous ecdysone deficiency in the above-mentioned larvae by an administration of the exogenous ecdysone.

C. THE EFFECT OF EXOGENOUS ECDYSONE ON THE ABILITY OF *D. VIRILIS* LARVAE TO METAMORPHOSE UNDER HEAT STRESS

Larvae unable to metamorphose under the effect of heat stress, synchronized by the second molt, were supplied with exogenous ecdysone, i.e., an aqueous solution of ecdysone (45 μg/g body weight) was mixed with the larval food. A control was served food medium supplemented with water.

It was found that addition of exogenous ecdysone to the food of larval *D. virilis* at a control temperature (25°C) accelerates the onset of metamorphosis (Table 5) and does not affect the survival (Table 6). On the other hand, exogenous ecdysone supply to the larvae unable to metamorphose at 32°C does ot elicit metamorphosis as might be expected. More-

TABLE 6

The Effect of Exogenous Ecdysone on the Survival of *D. virilis* Unable to Metamorphose at 32°C

	25°C		32°C	
Group of larvae	Number of larvae	Survival (%)	Number of larvae	Survival (%)
Control	105	72 ± 4.4	122	2 ± 1.3
Ecdysone-treated	81	74 ± 4.9	125	0

over, all the above larvae died 48 to 52 h after the second molt, viz., earlier than did the control ones. It is noteworthy that ecdysone stimulates the epidermal cells of the larvae. The cuticle of dead larvae was more tanned than of the controls and it resembled the pupal one.

The observation that exogenous ecdysone supply cannot stimulate metamorphosis in larvae of *R. prolixus* under heat stress was published in 1955 by Wigglesworth.[6] Discussing Wigglesworth's results, O'Kasha[9] suggested that some other factors, besides PTTH absence and ecdysone deficiency, may be involved in hindering the *R. prolixus* larvae to metamorphose under heat stress.

Having in mind the results of Wigglesworth[6] and O'Kasha,[7-9] we supposed that the capacity of insects to undergo metamorphosis under heat stress might depend on the presence of the third hormone involved in insect metamorphosis, the juvenile hormone[24,25] (JH).

III. CHANGES IN THE JUVENILE HORMONE CONTENT UNDER HEAT STRESS

It is well known that in decapitated, neck-ligated, and diapausing larvae and pupae of different insect species the exogenous JH stimulates the molting processes.[26-31] The above studies have demonstrated that in the absence of PTTH, the JH stimulates the PG to ecdysone synthesis, but its effect on this gland is indirect: JH stimulates the fat body to secrete a factor activating the PG.[32]

As already pointed out in the preceding sections, in insects developing under heat stress the PTTH secretion was inhibited.

All the data available allowed us to suggest that the ability of insects to undergo metamorphosis under heat stress may be due to the changes in the dynamics of the decrease in JH titer at high temperature. To verify this suggestion, it was necessary to estimate the JH content in larvae of *Drosophila virilis* able and/or unable to metamorphose when heat-stressed.

A. BIOASSAYS FOR MEASURING JH CONTENT IN *D. VIRILIS* LARVAE

Two bioassays were used to determine JH content in the larvae of *D. virilis*.

1. Estimation of JH Content Based on the Size of Larval Imaginal Disks

In *D. melanogaster* the size of the imaginal disks increases evenly during all the larval instars, somewhat slowing down during molts.[33] It is therefore unlikely whether the development of imaginal disks may be controlled by ecdysone or PTTH because the titers of these hormones show peaks just before molt.[34,35] In his studies on the growth of the implanted imaginal disks into adults of *D. melanogaster* Bodenstein[36] observed growth of these implants only in females. Experiments of Hadorn et al.[37,38] concerning implantation of imaginal disks into *D. melanogaster* males, and mated and/or virgin females demonstrated the greatest growth of the implants in the mated females, the smaller one in virgin females, and none

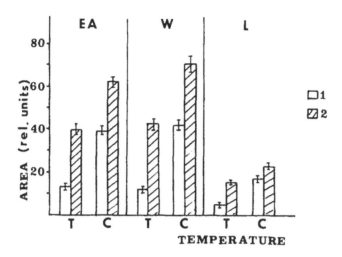

FIGURE 2. Effect of a juvenile hormone analog, ZR-512, on imaginal disk size in *D. virilis* larvae of line 147 under different temperatures. 1—control, 2—ZR-512-treated. T—32°C, C—25°C. Imaginal disks: EA—eye-antennal, W—wing, L—leg.

in males. Bührlen et al.[39] have measured the content of JH in *D. hydei* males and females. The JH content in 5-d-old females was almost threefold higher (2.45 fmol/individual) than in the males (0.87 fmol/individual); in 10-d-old females it amounted to 2.64 fmol, and in the males 1.71 fmol. Hadorn and Garcia-Bellido[38] have implanted imaginal disks into 4-d-old males and females of *D. melanogaster* and removed them after 8 d for size measurements. Their results (the greatest implant growth in females) provide evidence that the growth of imaginal disks in *D. melanogaster* is controlled by JH.

It should be noted that the presence of ecdysone in adult insects has led some authors[40,41] to the conclusion that growth of the imaginal disks implanted into *Drosophila* adults is due to the effect of ecdysone. However, it has been established that *D. melanogaster* males and females do not differ in ecdysteroid titer.[42] If the above-mentioned supposition that the imaginal disk growth is controlled by ecdysone holds true, there would not be differences in implant growth between males and females of *D. melanogaster*, as stated previously. Schwartz et al.[43] believed that the growth of imaginal disks is controlled by ecdysone, although their results did not support this possibility. They have demonstrated that ecdysterone content in the ecdysteroid-deficient mutant of *D. melanogaster* was twofold lower during pupariation than in larvae of the Oregon-R line, while the size of the imaginal disks was at this time much greater in the former than in the latter.

Direct experimental evidence that the growth of imaginal disks is under the control of JH was provided by *in vitro* studies of the above process on *D. melanogaster*[44] and *Manduca sexta*[45] by mixing the JH with the culture medium, as well as *in vivo* on *Laspeyresia pomonella*[46] larvae, treated with exogenous JH.

The aim of our experiments on the *D. virilis* wandering larval stage was to demonstrate the effect of JH on imaginal disk growth. The sizes of the imaginal disks in larvae that developed on food with and/or without a JH analog (45 μg/g body weight) at 25 and/or 32°C are presented in Figure 2. The results clearly indicate that at both temperatures the imaginal disks were much larger in the JH analog-treated larvae than in the untreated controls.

Consequently, *it is reasonable to estimate the relative JH content in the larvae of an identical developmental stage according to the sizes of their imaginal disks.*

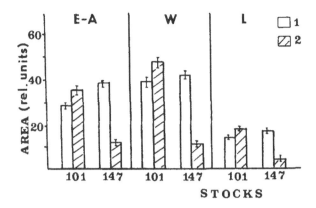

FIGURE 3.   Effect of high temperature on imaginal disk size in
D. virilis larvae of lines 101 and 147. 1—control (25°C), 2—heat-
exposed (32°C). Imaginal disks: EA—eye-antennal, W—wing, L—
leg.

2. Estimation of JH Content in the Larval Hemolymph in Relation to the Recipient's Rate of Development

It was established that injections of JH and/or CA extracts into wandering insect larvae delay or block larval-pupal molt,[47-49] and that 24 h before the onset of larval wandering the titer of JH is high,[39] whereas that of ecdysteroids is extremely low.[34] Hence, the hemolymph taken from *D. virilis* larvae 48 to 50 h after the second molt, i.e., 22 to 24 h before the larval wandering stage, and injected into wandering larvae would delay the metamorphosis of the recipients in a degree depending on the JH content in the hemolymph injected. Thus, according to the developmental rates of recipients one may conclude about the relative JH content in the hemolymph of donors.

B. EFFECT OF HEAT STRESS ON JH CONTENT IN *D. VIRILIS* LARVAE

The relative content of JH in the wandering, heat-stressed, and control larvae was estimated using a bioassay based on imaginal disks measurements. Concerning the above-mentioned bioassay, the results obtained (Figure 3) show that JH content was sharply decreased in heat-stressed larvae unable to metamorphose in comparison with their controls (line 147), while it was increased in those able to metamorphose (line 101) under the above experimental conditions.

To find out at what time during the heat-stressed third larval instar of line 147 the JH content starts to decrease, i.e., the growth of their imaginal disks slows down; we measured the imaginal disks in heat-treated and control larvae, 5, 24, and 48 h after the second molt.

During the first day after the second molt the content of JH was the same in the heat-treated and the control larvae (Figure 4), but after the following 24 h, it started to decline in the heat-stressed larvae as demonstrated by a slowed down disk growth. In addition, we also determined the content of JH in heat-stressed and control larvae of both lines, 22 to 24 h before wandering occurred. The relative JH content was estimated in relation to the rate of development of the recipient larvae injected with the hemolymph mentioned earlier.

The summarized results (Table 7) show that an injection of JH-1 into larvae delayed the onset of larval-pupal molt in the recipients. Hemolymph of control larvae of both lines and those heat-treated of line 101 postponed the onset of metamorphosis, too. This suggests that the JH content in the hemolymph was high sufficiently to delay metamorphosis. On the other hand, the JH content in heat-treated larvae of line 147 was very decreased owing to the fact that their hemolymph had practically no effect on the onset of metamorphosis. It must be mentioned that in the group of recipient larvae of line 147 injected with the

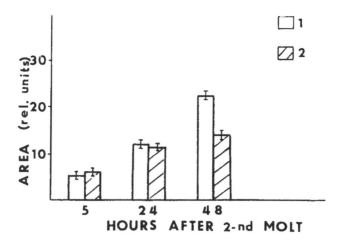

FIGURE 4. Effect of high temperature on the growth of the eye-antennal imaginal disk in *D. virilis* larvae of line 147. 1—control (25°C), 2—heat-exposed (32°C).

hemolymph of the heat-treated larvae of the same line, a number of individuals pupated after the pupation of all the larvae receiving Beadle medium was completed (31 h after the injection). These results show that in a small number of heat-treated larvae of line 147, 48 to 50 h after the second molt the JH content in the hemolymph was high enough to evoke the above-mentioned responses. This observation will be referred to later.

Hence, the results of these experiments indicate that during the development of *D. virilis* under heat stress JH content of wandering larvae was increased in larvae able to undergo metamorphosis, and it was sharply decreased in the unable ones. In the latter, the decrease in JH content starts 24 h after the second molt, i.e., 2 d earlier than in the controls.

Thus, it was shown that the content of ecdysteroids and JH was decreased in larvae unable to metamorphose under heat stress. Moreover, as reported earlier, the exogenous ecdysone supply was not capable of stimulating these larvae to undergo metamorphosis under heat stress. We attempted to compensate for the deficiency in JH in these larvae by treating them with a JH analog, the ZR-512.

C. EFFECT OF A JH ANALOG ON THE ABILITY OF *D. VIRILIS* LARVAE TO METAMORPHOSE UNDER HEAT STRESS

Larvae unable to metamorphose under heat stress (line 147) were treated with JH analog ZR-512. A solution of ZR-512 in acetone was mixed with the food medium. After evaporating acetone, larvae synchronized by the second molt were placed on the food (the amount of ZR-512 was 45 μg/g body weight). In the control cultures the food was mixed with the appropriate volumes of acetone which was evaporated.

It was found that treatment of the larvae with the JH analog at control temperature (25°C) considerably delays the onset of metamorphosis (Table 8) and that has no effect on the survival (Table (9). Treatment of the larvae with the JH analog at high temperature (32°) renders the larvae of line 147 able to metamorphose, when heat stressed (Table 9).

In such a way we have obtained confirmation of our supposition that in heat-stressed larvae when the PTTH secretion is blocked, precisely the presence of the JH enables the heat-stressed larvae to undergo metamorphosis. Our results revealed that in the above processes the basic role probably plays, the specific dynamics of the JH decrease just before the pupariation, since in the heat-stressed wandering larvae able to metamorphose the JH content was increased, while in those unable to metamorphose it was sharply decreased.

Having in mind that the above features were obtained in studies carried out on insects

TABLE 7

Effect of Hemolymph Injected to Heat-Exposed and Control Larvae of Lines 101 and 147 on Developmental Time of the Recipients

Line	Treatment	Number of larvae	Pupation time (hours after injection)											Dead larvae (%)
			11	13	15	17	21	27	32	35	40	45	50	
			Larvae pupated during time interval (%)											
101	I	28	14	32	32	22	0	0	0	0	0	0	0	0
	II	37	0	16	19	38	19	0	0	0	0	0	0	8
	III	30	0	0	0	17	27	17	20	3	3	0	0	13
	IV	30	0	0	0	7	23	23	20	7	7	3	3	7
	V	28	0	0	0	10.5	17.5	29	29	3.5	3.5	3.5	0	3.5
147	I	17	12	12	18	35	23	0	0	0	0	0	0	0
	II	25	0	12	16	24	20	20	0	0	0	0	0	8
	III	28	0	0	0	11	11	11	25	14	14	3	0	11
	IV	31	0	6.5	13	23	29	16	6.5	0	0	0	0	6

Larvae: I—intact controls; II—injected with Beadle medium (0.1, 1); III—injected with the hemolymph of control larvae (0.1, 1); IV—injected with the hemolymph of heat-exposed larvae (0.1, 1); V—injected with JH-1 (0.03 μg/mg of larval body weight).

TABLE 8

Effect of ZR-512, a JH Analog, on the Developmental Time of Larvae at 25°C

Group of larvae	Number of larvae	Pupation time (hours after second molt)					
		48	72	96	120	144	168
		Larvae pupated during time interval (%)					
Control	62	0	17	60	23	0	0
Treated with ZR-512	47	0	0	40	30	23	7

TABLE 9

Effect of ZR-512, a JH Analog, on the Survival of *D. virilis* Larvae Unable to Metamorphose at 32°C

Group of larvae	25°C		32°C	
	Number of larvae	Survival (%)	Number of larvae	Survival (%)
Control	102	74 ± 4.2	122	5 ± 1.4
Treated with ZR-512	77	81 ± 4.5	121	45 ± 4.7

of genetically selected lines, these features must be genetically determined and, consequently, there must exist a gene system controlling the ability of *D. virilis* larvae to metamorphose under heat stress.

The dynamics of the JH decrease before the pupariation processes is a result of relation between the rate of synthesis and the rate of degradation of the JH.

In the literature available to use we have not found publications concerning the allatotropic activity of the brain and the JH synthesis in insects developing under stressful temperatures. Our results show that differences in the JH content in *D. virilis* larvae able to metamorphose under heat stress are probably not related to the JH synthesis, since the physiological state of their CA under heat stress was similar to that observed in the controls (Figure 1). On the other hand, in larvae unable to metamorphose when heat-stressed, the CA are more active than those in the control larvae since the volume of nucleoli in the CA cells was increased (unpublished results).

The results of our detailed studies concerning the degradation of JH in larvae of *D. virilis* under heat stress will be described in the following section.

IV. DEGRADATION OF JH IN *D. VIRILIS* LARVAE UNDER HEAT STRESS

The major pathways for JH degradation in insects is hydrolysis of the ester moiety and hydration of the epoxy ring. The primary products of these reactions are JH-acid and JH-diol, respectively.[50]

In insect hemolymph, JH is degraded only through hydrolysis of the ester bound by a specific group of esterases. These JH-esterases are capable of hydrolyzing the JH, bound or unbound, to the carrier protein.[51] The other tissues contain, besides esterases (JH-esterases and "general" esterases), epoxyhydrases and mixed function oxydases capable of degrading unbound JH.[52] It should be noted that the general esterases show little[51] or no hydrolytic activity[53,54] with respect to JH.

It was established that JH-esterase plays the key role in the metamorphosis of insects of the order Lepidoptera, since it inactivates the JH in the hemolymph before reprogramming

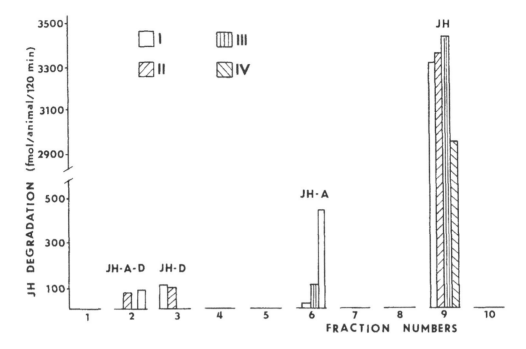

FIGURE 5.   Analysis of the degradation products of JH by thin layer chromatography. I, II, III, IV—JH degradation after incubation with larval extracts 48, 60, and 70 h after the second molt and with pupal extracts 48 h after pupariation, respectively. 1 to 10 designate the fraction numbers on the thin layer chromatograms. JH—juvenile hormone nondegradated, JH-A—JH-acid, JH-D—JH-diol, JH-D-A—JH-diol-acid.

of the genome from the larval to the pupal program of development and in switching to a new program.[35,45,55-58]

Studies of JH-esterase activity in the hemolymph of some species of the order Diptera, e.g., *Drosophila melanogaster,*[59] *Sarcophaga bullata,*[60] and *D. hydei,*[61] raised the question whether the JH-esterase plays the key role in the metamorphosis of Diptera since the JH-esterase activity was undetectable in the hemolymph of the last instar larvae. Activity of JH-esterase in the hemolymph started to appear in the prepupal stage.[61]

However, the absence of JH-esterase activity from the hemolymph of dipteran larvae did not mean that JH-esterase does not play the key role in their metamorphosis. It is possible that the degradation of JH occurs in the tissues of wandering dipteran larvae and not in their hemolymph. To see whether this is the case, measurements of JH-esterase activity in the whole larval body were needed.

A. ROLE OF JH-ESTERASE IN THE METAMORPHOSIS OF *DROSOPHILA*

The metabolism of ³H-JH-III by the supernatants of the larval and pupal homogenates of *D. virilis* was studied using a modification[53] of the method of Hammock and Sparks.[62] Thin layer chromatography (TLC) was performed according to Renucci.[63]

Figure 5 represents the results of a typical TLC separation of samples of JH degradation by the supernatants of larval (48, 60, and 70 h after the second molt) and pupal (48 h after pupariation) homogenates.

The results from TLC for the mid-third instar larvae (48 h after the second molt) are clear. JH is degraded by epoxyhydrase through hydration of the epoxy ring, since only JH-diol is chromatographically separated. Our results are in agreement with those obtained on larvae of two other dipteran species, *S. bullata*[50] and *D. melanogaster,*[59] but they are at variance with those of Klages and Emmerich[61] who have demonstrated the presence of JH-acid in the peripheral tissues of third instar larvae of *D. hydei.* This may be explained by

TABLE 10
Activity of JH-Esterase during
Development of *D. virilis*

Groups	JH-esterase activity (fmol/animal)
Larvae	
48 h after second molt	0
12 h before pupariation	0.07 ± 0.01
2 h before pupariation	0.41 ± 0.05
White prepupae	0.62 ± 0.1
Pupae	
24 h after pupariation	1.15 ± 0.15
48 h after pupariation	1.8 ± 0.2
77 h after pupariation	1.18 ± 0.2
90 h after pupariation	1.15 ± 0.2

some differences in the methods used to measure the degradation of ^{3}H-JH, i.e., different substrates were used. In our case, it was JH-III, the natural *Drosophila* hormone, while Klages and Emmerich[61] used JH-1.

Twelve hours before pupariation, i.e., 60 h after the second molt, JH is degraded mainly by epoxyhydrase, but JH-esterase activity was also stated since the TLC detected some amounts of JH-acid (Figure 5).

At the end of the wandering larval stage (70 h after the second molt) JH was degraded soley by the JH-esterase, as shown by the presence of the JH-acid alone on the chromatograms (Figure 5).

In pupae, 48h after pupariation, the activity of JH-esterase is high (the major product of JH degradation is JH-acid), but the epoxyhydrase was active too, as estimated according to the negligible amounts of JH-diol-acid separated by TLC (Figure 5).

The values for JH-esterase activity during development of *D. virilis* are given in Table 10. In larvae of the mid-third instar, JH-esterase activity was undetectable. At the beginning of the wandering stage (12 h before pupariation) it was low, but at the end of this stage it sharply rose. After pupariation, JH-esterase activity continues to rise attaining the peak values in 48-h-old pupae; thereafter it starts to decline.

On the basis of our observations, it was concluded that in the wandering larvae of *Drosophila* JH was degraded by JH-esterase directly in the larval tissues and not in the hemolymph, as demonstrated in Lepidoptera.

The conclusion that JH-esterase plays the key role in processes of metamorphosis in Lepidoptera has been drawn from a comparison of the dynamics of JH-esterase activity, JH level, and ecdysteroid titer: an increase in JH-esterase activity and a decrease in JH titer precede the peaks in the level of the ecdysteroids responsible for reprogramming of the genome and for the switching on the new program.[64-66]

From this point of view, it would be interesting to compare the ecdysteroid titer, the decrease in JH level, and increase in JH-esterase activity in *Drosophila*.

In studies carried out on *D. melanogaster* in the course of several years and systematized by Richards,[34] it was demonstrated that ecdysteroids showed a peak either during puparium formation[67-69] or an increase in the hormone titer 5 to 6 h before pupariation, with the highest values 3 to 4 h following puparium formation.[70,71] Moreover, common to all the above studies was an intermolt level of ecdysteroids, 7 to 8 h before pupariation. Thus, in *D. melanogaster* an increase in ecdysteroid titer is linked to the metamorphosis. It starts at the wandering stage which lasts 6 to 10 h,[35] while the major peak in the ecdysteroid titer is reached during puparium formation (first peak).

Our data indicate that in *D. virilis* it is precisely at the time of wandering that JH-esterase activity rises sharply, approximatively sixfold. This rise is well correlated with the decrease in the JH titer reported for another species of *Drosophila*, of the *virilis* group, the *D. hydei*.[39]

It should be emphasized that the events involved in the reprogramming from a larval to a pupal development are not limited to pupariation. At the prepupal and early pupal stages of *Drosophila* a sequence of inductive interactions takes place, as evidenced by the appearance of a second ecdysteroid peak in prepupae[42,69] and a third one in early pupae.[42]

It is noteworthy that in Lepidoptera the second peak of ecdysteroid titer related to the switchover of the developmental program also occurs at the prepupal stage.[72] Furthermore, Jones et al.[72] have observed that in the prepupae of *Trichoplusia ni* treated with a JH-esterase inhibitor the ecdysterone titer does not decrease to a control level and the pupal ecdysis is blocked. Provided that ecdysteroids have to exert their inductive effects on the prepupae of Diptera, JH titer has to be extremely low. The above was confirmed in the studies of Richards,[73] who has established that the presence of JH at an ecdysone-free period (at the prepupal stage between the first and second peaks of ecdysteroid titers) suppresses the formation of ecdysone-induced puffs of polytene chromosomes in the cells of *D. melanogaster* salivary glands. Lezzi and Wyss[74] have shown that the presence of JH blocks the formation of the ecdysone-induced puff on locus 1-18-C of salivary gland chromosomes of *Chironomus* prepupae.

In fact, the data of Bührlen et al.[39] showed an extremely low JH content during the prepupal stage of *D. hydei*, while our results demonstrated that JH-esterase activity was high in the prepupae of *D. virilis*. High JH-esterase activity in the prepupae and early pupae of *D. hydei* was reported by Klages and Emmerich.[61] This high JH-esterase activity during the above-mentioned developmental stages is probably related to the need for an intensive JH metabolism at these periods.

In his extensive review concerning the mechanisms of JH degradation, Hammock[57] supposed that JH may be necessary at the prepupal stage as a synchronizer of development by slowing down the differentiation of tissues particularly sensitive to the effect of ecdysone and/or to the stimulation of ecdysone synthesis. Afterwards, Newitt and Hammock[58] concluded that JH did not exhibit a prothoracicotropic effect on the prepupal PG. Nevertheless, the necessity of JH presence at the prepupal stage is obvious. In fact, in many lepidopteran species there is the JH pulse in the prepupae.[29,75-78] Undoubtedly, the measurement of JH titer in *Drosophila* prepupae at shorter time intervals than those of Bührlen et al.[39] might reveal the existence of the prepupal JH pulse also in this species.

However, as noted above, the presence of JH in the prepupae blocks the switchover of the developmental program in the target tissues. Accordingly, one would anticipate a high metabolic rate of JH in prepupae and possibly in the early pupae: JH synthesis and its rapid degradation by the JH-esterase in the hemolymph. It is noteworthy that at the prepupal stage JH-esterase activity is regulated through direct induction by JH.[79]

In summary, the role of JH-esterase in the switchover of the developmental program in the metamorphosis of Diptera may be demonstrated on *Drosophila*. The decrease in JH titer, necessary for the larval-pupal reprogramming of the genome of tissue cells to be the first involved in the processes of metamorphosis, occurs in wandering *Drosophila* larvae through the action of JH-esterase which hydrolyzes JH directly in the larval tissues. It must be mentioned that in *Drosophila* the reprogramming may not proceed simultaneously, as shown by Riddiford[47,80] in *Hyalophora cecropia*, in which the epidermal cells are reprogrammed first, followed by those of the gonads, muscles, and digestive tract. The further decrease in the JH titer or its maintenance at low levels, a prerequisite needed for the switchover of the developmental program related to pupation, is caused by its hydrolysis by JH-esterase in the hemolymph of prepupae or early pupae. The relatively high activity of JH-esterase during pupal development remains high until the JH degradation is completed, i.e., a condition necessary for the initiation of the second metamorphosic molt.[35,55-57] Thus, JH-esterase represents the major link in the processes of JH degradation during the metamorphosis of *D. virilis*. Hence, the genetically determined differences in the dynamics of JH inactivation in larvae able or unable to metamorphose under heat stress should be due to the differences in the time of JH-esterase activation in these larvae, while the gene system regulating JH-esterase activity should control this process.

B. CHANGES IN JH-ESTERASE ACTIVITY IN *D. VIRILIS* UNDER HEAT STRESS

To study the structure of gene system controlling JH-esterase activity and its function under heat stress, the enzyme has to be assayed in single individuals. The above-mentioned method of measuring JH-esterase activity, based on the hydrolysis of labeled JH, does not make possible JH assays in individual *Drosophila* larvae because of their small size.

For an analysis of JH-esterase activity in individual *Drosophila* the technique of electrophoresis with subsequent identification of enzyme activity by use of an artificial substrate offers good results, provided that one knows which particular fraction in the esterase patterns represents the JH-esterase.

1. Identification of JH-Esterase in the Esterase Patterns of *D. virilis*

The identification of JH-esterase in the esterase patterns has been previously described in detail.[54] Some of the key experiments that allowed us to detect JH-esterase fractions will be described below.

After the definition of Hammock[57] the JH-esterase in insects is an enzyme with high affinity for JH, capable of hydrolyzing either the unbound or the bound JH to the carrier-proteins. Its activity increases in response to the exogenous JH. The enzyme is not inhibited with the specific inhibitor of the "general" esterases, the diisopropylphosphofluoridate (DFP). Located in the larval fat body, the changes in its activity are correlated with the dynamics of changes in JH titer during insect development.

The significant findings of our experiments were as follows:

1. In a substrate for fermentation reaction of the mixture of JH with β-naphthylacetate, JH is a substrate only for a single fraction in the *D. virilis* esterase patterns, the pupal (p)-esterase. The JH inhibited the activity of only p-esterase (Figure 6) and showed an inhibition of the competitive type (Figure 7).
2. The Michaelis constant value of p-esterase for β-naphthylacetate was close to the one obtained for JH.[54]
3. DFP inhibited all the fractions in the *D. virilis* patterns with the exception of p-esterase.[54]
4. Exogenous JH activated p-esterase in *D. virilis* prepupae.[25]
5. The dynamics of changes in p-esterase activity during development of *D. virilis* was identical to that of JH-esterase (Figure 8).
6. p-Esterase is located in the fat body of larval *D. virilis*.[25]

The above characterization of *D. virilis* p-esterase is consistent with that given for JH-esterase by Hammock.[57] Thus *D. virilis* p-esterase is a JH-esterase, and an electrophoretic analysis may be well applied for determination of its relative activity in studies of heat stress effect(s) on JH-esterase activity.

2. Activity of JH-Esterase in *D. virilis* under Heat Stress

JH-esterase activity evaluated according to the staining intensity of the corresponding fraction in the esterase patterns was compared in heat-exposed and control pupae, able (line 101) and unable (line 147) to metamorphose when heat-stressed. The pupae from cultures exposed to 32°C during the last larval instar will be henceforth referred to as heat-treated.

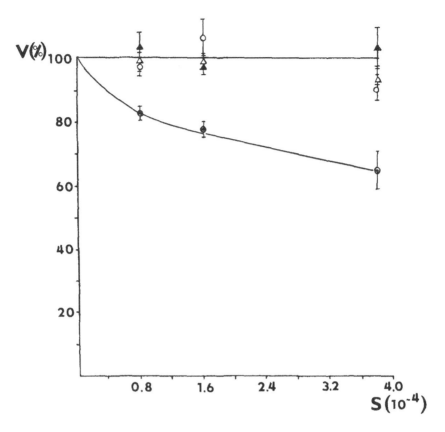

FIGURE 6.   Dependence of the velocity of the hydrolysis of β-naphthyl acetate (concentration 2.7×10^{-4} M) on JH-1 added to the reaction mixture. V—reaction velocity, S—JH-1 concentration. Open circles, esterase-2; filled circles, p-esterase. Open triangles, esterase-4; filled triangles, esterase-6.

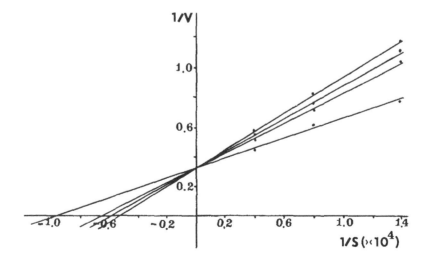

FIGURE 7.   Inhibition of p-esterase activity by JH-1 (Lineweaver-Burk plot). V—reaction velocity, S—concentration of β-naphthyl acetate. 1—reaction without the addition of JH-1; 2—0.8×10^{-4} M JH-1 added; 3—1.2×10^{-4} M JH-1 added; 4—1.6×10^{-4} M JH-1 added.

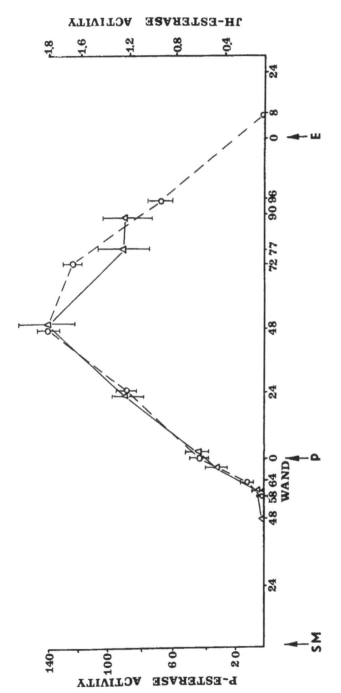

FIGURE 8. Changes in the activities of p-esterase (rel. units/animal) and JH-esterase (fmol/min/animal) during the development of *D. virilis*. Circles — JH-esterase, means ± SD (5 to 7 independent extracts, determinations in triplicate); triangles — p-esterase, means ± SD (11 independent extracts). SM — second molt, P — pupariation, E — emergence, WAND — wandering.

It was found that JH-esterase was not active in the heat-treated pupae of line 101, while in those of line 147, the enzyme was active. Afterward we determined the activation time of JH-esterase during the third larval instar under heat stress in larvae able and those unable to metamorphose under these conditions. The data summarized in Table 11 demonstrate that JH-esterase is activated 48 h earlier in the larvae unable to metamorphose under heat stress than in the controls. Thus, from the study of JH-esterase activity in developing larvae, able and/or unable to metamorphose when heat-stressed, one may conclude that the differences in JH content in the above larvae might be due to the different level of JH-esterase activity. As a matter of fact, the elevated JH content in the wandering larvae able to metamorphose under heat stress is explained by the inactivity of JH-esterase. At this time, the decrease in JH content in larvae unable to metamorphose when heat-stressed is caused by an early activation of JH-esterase. It is necessary to recall that some increase in CA activity in larvae unable to undergo metamorphosis under heat stress (Section III.C) is evidently in connection with an early activation of JH-esterase: an abrupt decline in JH titer is, in these larvae, compensated by an activation of JH synthesis. It is worthy to note that 48 h after the second molt, in 10% of larvae (line 147) JH-esterase activity is not yet detected and, as already indicated in Section III.B, in about 10% of these larvae JH content is not yet reduced.

To recapitulate, changes in JH content in heat-stressed *D. virilis* larvae, after a delay in development able to metamorphose under heat stress, are controlled by changes in the activity of JH-esterase. In this regard, Diptera differ from Lepidoptera in which, under thermal stress, the JH titer is controlled not only through a decrease in JH-esterase activity, but also through a substantial increase in the activity of CA.[81,82] It is not excluded that besides differences between the above orders related to the fixed number of molts in *Drosophila* and the possible indefinite number of molts in Lepidoptera, there are some differences in the mechanisms of the heat and cold stress (see the literature cited on the effect of cold stress on *Galleria mellonella*).

As indicated in Section III.C, differences in the activity of JH-esterase and, consequently, in JH content in heat-stressed *D. virilis* larvae should be genetically determined, and there should be a gene system controlling the ability of *D. virilis* larvae to metamorphose under heat stress. This genetic system will be described in the next section.

V. THE GENETIC SYSTEM CONTROLLING JH-ESTERASE ACTIVITY AND THE ABILITY TO METAMORPHOSE UNDER HEAT STRESS IN *D. VIRILIS*

The gene system controlling JH-esterase activity and thereby the ability of *Drosophila virilis* to metamorphose under heat stress has been described in detail earlier.[83-85] Only the key experiments which allowed us to determine the structure and function of this system under optimum conditions and under heat stress are reviewed in the following section.

A. INHERITANCE OF THE ACTIVITY LEVEL OF JH-ESTERASE AND THE ABILITY TO METAMORPHOSE UNDER HEAT STRESS IN *D. VIRILIS*

D. virilis individuals, able to metamorphose under heat stress (line 101) and unable to do so (line 147), were used in the genetic experiments.

The F_1 hybrids between lines 101 and 147 metamorphosed and survived under conditions of heat stress. The activity of JH-esterase was sharply decreased in the hybrid pupae under the above conditions.[83] The results of the above-mentioned cross-breeding tests demonstrated that both the survival rate and the activity of JH-esterase are under the control of a single gene in heat-stressed *D. virilis* (Tables 12 and 13). These results support the conclusion reached in studies concerning the endocrine control of development in *D. virilis* under heat stress, that the ability to metamorphose under these conditions is due to the activity level

TABLE 11

The Time of Activation of JH-Esterase (JHE) in *D. virilis* Larvae, Able and Unable to Metamorphose under Heat Stress

	Hours after the second molt							
	24		48		72		96[a]	
Group of larvae	Number of larvae	Larvae with active JHE(%)	Number of larvae	Larvae with active JHE(%)	Number of larvae	Larvae with active JHE (%)	Number of larvae	Larvae with active JHE(%)
Line 101 (25°C)	—	—	90	0	85	87 ± 3.7	—	—
Line 101 (32°C)	—	—	90	0	84	0	—	—
Line 147 (25°C)	—	—	75	0	106	84 ± 3.6	—	—
Line 147 (32°C)	86	43 ± 5.4	119	81.4 ± 3.6	90	88 ± 3.4	89	95.6 ± 2.2

[a] All the line 101 larvae and line 147 controls have pupated by 96 h after the second molt.

TABLE 12
Segregation for the Activity Level of JH-Esterase
under Heat Stress in the Test Crosses Line 147 ×
(Line 147 × Line 101)F₁

	Number of individuals with an activity level		
Data	(Line 147 × line 101)F₁	Line 147	Total
Observed	75	69	144
Expected on the basis of a 1:1 ratio	72	72	144

TABLE 13
Segregation for Viability under Heat Stress in the Test
Crosses Line 147 × (Line 147 × Line 101)F₁

Data	Number of larvae	Number of individuals survived	Number of individuals dead
Observed	506	239	267
Observed, normalized[a]	506	244	262
Expected on the basis of a 1:1 ratio	506	253	253

[a] Survival was not 100% for the control larvae due to heat-unrelated causes. For this reason, their survival was taken as 100% and survival for the heat-exposed larvae was normalized accordingly.

of JH-esterase. However, it is important to note that this conclusion holds true only in the case when other systems promoting the metamorphosis under thermal stress function normally under these conditions. All the mutations which prevent the normal activity of those systems are thermosensitively lethal since individuals with these mutations will not be able to undergo metamorphosis under heat stress.

Thus, it was established that under heat stress the low JH-esterase activity in individuals able to metamorphose and the high one in those unable to metamorphose are under the control of alleles of a single gene. These alleles at the structural locus (*Est-JH*) might be the code for the synthesis of JH-esterase and they might determine the differences in the primary structure of the enzyme, affecting the efficiency of catalysis at high temperature, and particularly the decrease in JH-esterase activity in individuals of line 101.

Analysis of the dynamics of heat inactivation of JH-esterase in the extracts obtained from the pupae of lines 101 and 147 showed that the above assumptions were not exact: the enzyme of both lines was inactivated in the same manner when heat-exposed (Table 14). Thus, even an existence of differences in the primary structure of JH-esterase molecule between lines 101 and 147 may not be the cause for the almost complete absence of the enzyme activity in the individuals of line 101 under heat stress. Consequently, the results furnished an evidence for the existence of a gene, *Ra-JHE*, regulating the expression of the structural locus *Est-JH*. In *D. virilis*, the *Est-JH* locus has been located on chromosome II[86] and the Ra-JHE locus on chromosome VI.[84] Under optimal conditions, the Est-JHE gene is active in *D. virilis* larvae (at least from the second larval instar) determining the synthesis of the inactive form of JH-esterase in the fat body.[25]

Experiments designed to elucidate the mechanism of the Ra-JHE gene action have been carried out. These experiments will be briefly described in the next section.

TABLE 14
Dynamics of Heat Inactivation of JH-Esterase during Incubation of
48-h Pupal Extracts of Lines 101 and 147 at Different
Temperatures for 15 min and at 42°C for Different Times

Time (min)	Relative JH-esterase activity[a] (%)		Temperature (°C)	Relative JH-esterase activity[a] (%)	
	Line 101	Line 147		Line 101	Line 147
10	76 ± 5	73 ± 9	37	83 ± 6	79 ± 6
20	74 ± 4	60 ± 6	42	73 ± 5	70 ± 7
30	50 ± 5	50 ± 5	45	60 ± 3	58 ± 6
40	36 ± 3	42 ± 4	47	15 ± 3	16 ± 3
50	24 ± 3	26 ± 3	52	0	0
60	13 ± 1.5	10 ± 2			

[a] Relative JH-esterase activity is expressed as the ratio of the peak area of a JH-esterase fraction in a heat-exposed sample to that in the control sample (not incubated) on the densitogram.

TABLE 15
Relative JH-Esterase Activity in 48-h Pupae of Line 101
at Different Temperatures

Pupae	Number of replicates	Relative JH-esterase activity[a] (%)
Control	8	11.7 ± 0.27
Heat-exposed	8	0
Mixture of homogenates of control and heat-exposed	8	11.2 ± 0.3

[a] Relative JH-esterase activity is expressed as the ratio of the activity peak area of a JH-esterase fraction to the sum of the activity peak areas of all the other fractions of a sample on the densitogram.

B. THE MECHANISM OF ACTION OF THE GENE REGULATING JH-ESTERASE ACTIVITY IN *D. VIRILIS* UNDER OPTIMAL CONDITIONS AND HEAT STRESS

It was assumed that the Ra-JHE gene may control the synthesis (or secretion) of a factor stimulatory or inhibitory for JH-esterase. Simple experiments were performed to find out if the putative factor exists: JH-esterase activity was compared in the homogenates of heat-exposed and control pupae of line 101, and in a mixture of both the above-mentioned homogenates after 2 h incubation at room temperature (Table 15). JH-esterase activity in the homogenate mixture was found to be the same as in that of the control pupae. This demonstrates that the homogenates of the control pupae contain an activating factor (AF) for JH-esterase. Experiments with the incubation of the brain homogenates from control larvae of line 101 or those heat-exposed of line 147, whose brains were severed 4 h before pupariation, together with the homogenates of the heat-exposed pupae of line 101, demonstrated an activation of JH-esterase in the latter.[84] However, no activation of JH-esterase was observed when homogenates of the heat-exposed pupae of line 101 were incubated with the brain homogenates of heat-exposed larvae of line 101.[84] The enzyme was also activated when the homogenate from mid-third instar larvae which does not contain an active form of JH-esterase was incubated with the brain homogenate.[84] JH-esterase was not activated in analogous experiments in which the brain homogenates were exposed to 100°C or treated with pronase before incubation.[84]

The results of the above series of experiments allowed us to conclude:

1. Under control and stressful temperatures the *Ra-JHE* gene regulates the expression of the structural *Est-JH* locus at the posttranslational level controlling the synthesis (or secretion) of the AF which converts the inactive form of JH-esterase into active.
2. In larvae able to metamorphose when heat-stressed, the *Ra-JHE* gene is not expressed and AF is missing (this allele of the gene is referred to as *Ra-JHE* [1−].
3. In larvae, unable to metamorphose when heat-stressed, the *Ra-JHE* gene is expressed and AF is secreted (the *Ra-JHE* [1+] allele is heat-sensitive lethal, i.e., its carriers die when developing under heat stress).
4. The larval brain is the source of AF.
5. AF is of a proteinaceous nature.

The two last conclusions led us to compare the secretion of AF and the physiological state of the larval protocerebral NSCs under heat stress in *D. virilis* able and unable to metamorphose when heat-stressed. An evaluation of the physiological state of NSCs in the medial group, in which in *D. virilis* the PTTH is synthesized, showed some differences in the response of these cells to the effect of stressful temperature in larvae able and those unable to undergo metamorphosis. In larvae able to metamorphose under stressful temperature all the NSCs were inactive[13] and they did not produce AF under the above conditions. In those unable to metamorphose under heat stress, two NSCs in the medial group, comprising eight to ten NSCs in each brain lobe, located laterally, are active (Figure 9) and AF is synthesized. The above-mentioned correlation makes us able to suppose that the AF is synthesized in these cells.

VI. MECHANISM OF THE STRESS RESPONSE IN *D. VIRILIS* DEVELOPING UNDER HEAT STRESS

In summary, we propose the following scheme concerning the mechanism of the stress response in *Drosophila virilis* developing under heat stress (Figure 10).

Before turning to the endocrine control of development under heat stress, it would be worthwhile to recall how the switchover from a larval to a pupal program, a prerequisite for the onset of metamorphosis, takes place under optimum conditions. At the end of the last larval instar, the NSCs secrete PTTH which stimulates the PG to synthesize ecdysone, which produces a switchover of the developmental program provided that JH titer is sharply decreased. The decrease in JH titer is rapidly and efficiently effected by JH-esterase whose activity in *D. virilis* is controlled by a gene system consisting of at least two loci. At the end of the second instar, the structural locus *Est-JH*, determining the synthesis of the inactive form of the enzyme, starts to function in the fat body. At the end of the third (last) larval instar, the *Ra-JHE* gene regulating the function of the *Est-JH* locus at the posttranslational level is expressed. The *Ra-JHE* gene determines the synthesis (or release) of AF, converting the inactive form of JH-esterase into the active one. By degrading the JH, JH-esterase enables ecdysone to switch the developmental program over.

Heat stress inhibits the activity of NSCs and, as a consequence, the secretion of PTTH and AF ceases. The cessation of PTTH secretion delays metamorphosis, thereby giving the larvae the opportunity to survive the stressful conditions. In the case of prolonged stress the simulation of ecdysone synthesis is taken over by JH, whose degradation is slowed down because of the absence of the active form of JH-esterase. The JH inactivation needed for the program switchover is presumably more time-consuming, and in this case it is effected

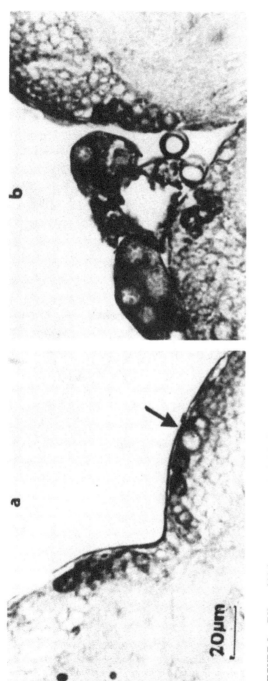

FIGURE 9. Effect of high temperatures on the NSC of *D. virilis* larvae. (a) The line 147 (the arrow points to the neurosecretory cell remaining in the active state); (b) the line 101.

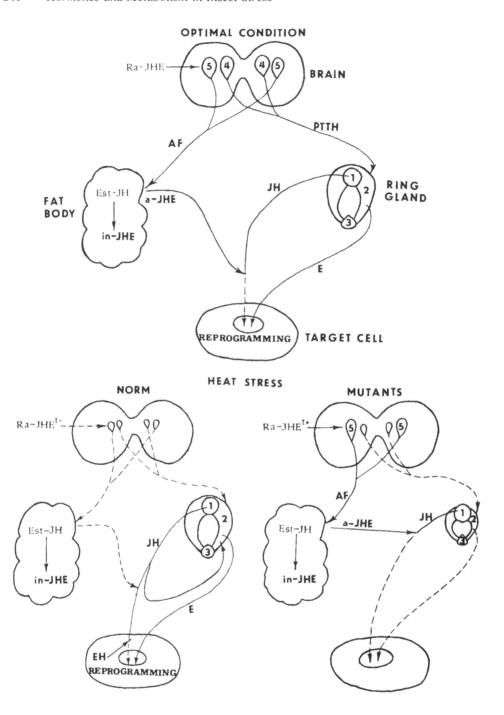

FIGURE 10.   Schematic diagram suggested for the stress response to high temperature in *D. virilis*. 1—*corpus allatum*, 2—peritracheal gland, 3—*corpus cardiacum*, 4—neurosecretory cells producing PTTH, 5—neurosecretory cells secreting AF. AF—activating factor, PTTH—prothoracicothropic hormone, E—ecdysone, JH—juvenile hormone, EH—epoxydhydrase, Est-JH—the structural locus of JH-esterase, Ra-JHE—the gene regulating the activity of JH-esterase, in JHE—inactive form of JH-esterase, a JHE—active form of JH-esterase.

by epoxyhydrase directly in the tissues. The structural *Est-JH* locus is active in all the heat-stressed larvae as it is in those maintained under optimum temperature conditions. In addition, the expression of the *Ra-JHE* gene is blocked by the stressful conditions and AF is not secreted into the hemolymph. This gives the heat-stressed *D. virilis* larvae the opportunity to undergo metamorphosis. However, in *D. virilis* populations carriers of the mutant alleles at the gene *Ra-JHE* are observed. These alleles express themselves under stressful conditions and, as a result, AF appears and converts the inactive JH-esterase into the active form. The latter degrades JH, the PG remains inactive, and ecdysone is not synthesized. Larvae with the mutant alleles are unable to pupate under heat stress and they die.

Is the mechanism promoting metamorphosis in *D. virilis* under heat stress operative under the given type of stress and in the given species only, or is it operative under the effect of other stressful environmental conditions and in other insect species, too?

To answer these questions we shall consider the links of the mechanisms, one by one.

A. INACTIVATION OF NSCs PRODUCING PTTH AND CESSATION OF ITS SECRETION

There is no doubt that the given response is observed in many insect species developing under stressful conditions. It has been described in Hemimetabola, *Rhodnius prolixus*,[2,6-8] and Holometabola, *Bombyx mori*,[87] *Lucilia sericata*,[88] *Sarcohpaga peregrina*,[89] *S. argiros-toma*,[90] and *D. virilis*[91,92] under the effect of various stressful external factors including high and low temperatures, high density, inadequate food supply, high humidity, etc. The same response has been described for insects in which stressful environmental stimuli induce the diapause.[14,26,93-99]

B. DELAY OR INHIBITION OF ECDYSONE SECRETION

In numerous studies devoted to development of different nondiapausing and diapausing insect species under unfavorable conditions, a delay or prevention of metamorphosis was observed, suggesting a delay or inhibition of ecdysone secretion in the above insects.[3-5,100-114] Concerning ecdysone secretion, the same was observed in experiments with the removal and implantation of PG into diapausing pupae of *Hyalophora cecropia*[26] and *Antheraea pernyi*[93] in response to the effect of low temperature and in larvae of *R. prolixus* under the effect of high temperature.[2,7] An inhibition of PG activity when metamorphosis was delayed under the effects of high density and inadequate food supply has been reported for *D. virilis*.[91,92] Berreur et al.[115] have compared development in *Calliphora erythrocephala* larvae maintained under low and high density conditions and have simultaneously measured their ecdysone titer. They found that when uncrowded larvae started to pupate, ecdysone titer was extremely low in the crowded ones; it started to rise on the next day and thereafter the crowded larvae started to pupate. Thus, it may be concluded that inhibited ecdysone secretion is a result of insect development under stressful conditions.

C. INCREASE IN JH CONTENT AND LOW ACTIVITY LEVEL OF JH-ESTERASE

In studies concerning hormonal control of larval diapause in *Diatraea grandiosella*, *Chilo partellus*, *C. suppressalis*, and *Ostrinia nubilalis*, it was reported that the JH titer was significantly higher in prediapausing and diapausing larvae than in the nondiapausing ones of an identical instar.[46,48,116-120] Mane and Chippendale[121] have demonstrated that the cause of elevated JH content in diapausing *D. grandiosella* larvae is the absence of JH-esterase activity. Bogus and Cymborowski[122] have observed that an exposure to 0°C brings about an elevated JH titer in *Galleria mellonella* larvae. Examining the delay of metamorphosis in *G. mellonella* induced either by low temperature or injury, McCaleb and Kumaran[123] have noted an absence of JH-esterase in the larvae. Hammock et al.[124] and Cymborowski et al.[125] have demonstrated a much lower activity of JH-esterase and a considerably higher JH titer in the last instar larvae of *Trichoplusia ni* and *Manduca sexta* before the onset of wandering (before development-reprogramming) as compared to the controls, when metamorphosis was delayed by starvation or rearing on a nutritionally deficient medium. Reddy et al.[126] have observed a decrease in JH-esterase activity in starved *G. mellonella* larvae. In addition, a decline in JH-esterase activity and an elevation of JH titer were observed in *G. mellonella* larvae developing under unfavorable photoperiod conditions.[127]

Furthermore, an increase in JH content and a decrease in JH-esterase activity were established in *Drosophila virilis* larvae when delay of metamorphosis occurred under the effect of crowding or after rearing on a deficient diet.[13,92,128] An elevation in JH content and a lowering in JH-esterase activity was demonstrated for the last larval instar of *G. mellonella* whose delayed metamorphosis was elicited by an injury of the integument.[129] It should be noted that the increase in JH titer in insects with delayed metamorphosis evoked by the effect of stressful factors is observed both in insects with fixed number of instars, as well as in those with an indefinite number of larval molts. In the first group, the function of CA is not changed under the effect of stressful extrinsic factors (heat, high density, deficient diet),[13,25] whereas in the second one, CA function is considerably enhanced under the effect of a stressor (cold).[81,82]

Thus, it is reasonable to conclude that the obtained hormonal responses in *D. virilis* larvae developing under thermal stress (inactivation of NSC producing PTTH and AF, delay or inhibition of ecdysone secretion, decrease in JH-esterase activity, and increase in JH content) are characteristic of other kinds of stresses and other insect species, too.

Quite plausibly, evolution has adopted two "survival strategies" for insects developing under unfavorable extrinsic conditions: (1) the facultative diapause and (2) the delayed or prevented metamorphosis (comprising the occurrence of supernumerary molts) in nondiapausing insects.[1] The two strategies insure insect survival under unfavorable conditions. The hormonal basis of both strategies comprises the above-mentioned reactions and, consequently, these responses are of an adaptive, protective nature. Hence, this hormonal response has two basic characteristics: unspecificity in relation to the given stressor (occurs under the effect of different stressors) and adaptivity. However, precisely these characteristics of the hormonal response are termed in the mammals as stress in its broad sense.[130] Both in mammals and insects, under stressful conditions there arises an unspecific standard hormonal response of an adaptive nature. The studies of Vigas[131] and Sudakov[132] have demonstrated that in the mammals the mechanism of stress response is unspecific, although there exists some degree of specificity in its expression: (1) the effect of a stressor depends on its intensity (there is a threshold level for the given stressor to act stressfully); and (2) the response is genotype-dependent (the same stressor can produce stress in some organisms and not in others[131,132].) Here, again, there is an obvious similarity between the response observed in our experiments on insects and those shown in mammals: (1) a threshold for the response to a particular stressor was demonstrated for *R. prolixus* and *D. virilis*. So the larval development of *R. prolixus* at 32°C does not affect the metamorphosis into adults, while a temperature of 35°C delays the molting into adults. In *D. virilis* a density of 40 larvae/vial is without effect on metamorphosis, but that of 80 larvae/vial postpones its onset;[128] (2) genotype dependency follows from the data indicating that a temperature of 32°C produces a stress response in *D. virilis*,[13] while in *R. prolixus*[2,6] this temperature is at the subthreshold level to act as a stressor.

Furthermore, it has been demonstrated that the same environmental cue can be stressful or not for the given species, depending upon the developmental stage of the insect at which it acts. For example, crowding is not consistently unfavorable for insect development. Depending on the life pattern of a species, low population density can cause stress, too. Corroborative are the studies of Singh and Pandey[133] carried out on *Diacrisia obliqua* larvae of different instars. The females lay eggs in heaps on leaves and the hatched larvae remain in close contact during the first and second instars. However, at the beginning of the third instar, the larvae start to migrate and spread over different plants. Singh and Pandey[133] have reared first and third instar larvae in petri dishes (diameter 15 cm) at a density of 1, 5, 10, 15, 20, and 25 larvae per dish. Developmental rates of larvae of the two instars at high and low densities showed that the longest development and the highest mortality were observed in cultures of the first instar larvae at a low density, as well as in those of the third instar larvae at a high density. Consequently, low density was stressful for developing first instar larvae, and high density for the third instar larvae. These results were not unexpected when the life pattern of this particular species was taken into account. Very illustrative are the studies of Ivanović et al.[10,11] who have demonstrated that there are seasonal differences in the response of *Morimus funereus* larvae to the effect of different temperatures. A temperature of 8°C was stressful to the larvae collected in June (inhibition of NSC secreting PTTH), but on those collected in November it was not stressful. In contrast, a temperature of 23°C provoked a stress in November, but not in June.[10,11]

It should be noted that prolonged exposure to a stressor of high intensity can evoke accelerated development, i.e., precocious metamorphosis in insects. Thus, in his experimental series, Chernysh[134] has demonstrated that immersion of *Calliphora erythrocephala* larvae into 40% formol for 180 min sped up puparium formation and that for 90 min slowed it down. An acceleration of metamorphosis occurs in electroshocked *B. mori* larvae (cited from Chernysh[134]). Precocious metamorphosis has also been observed in larvae with a cuticle injury made by the ovipositor of parasitic females.[135] The above effect occurred only after numerous injuries of the cuticle, i.e., a single puncture did not initiate precocious metamorphosis. Moderate injury (threshold) of the cuticle elicits a standard stress response involving a delay of metamorphosis, an inhibition of PTTH secretion, a delay of ecdysone secretion, an increase in JH content, and a decrease in JH-esterase activity.[7,8,13,123,129]

ACKNOWLEDGMENT

The author is grateful to Mrs. A. Fadeeva and Dr. V. Stanić for translating the review from Russian into English.

REFERENCES

1. **Novak, V. J. A.,** *Insect Hormones,* Chapman and Hall, London, 1975.
2. **Wigglesworth, V. B.,** Hormone balance and the control of metamorphosis in *Rhodnius prolixus* (Hemiptera), *J. Exp. Biol.,* 29, 620, 1952.
3. **Mellanby, K.,** Acclimatization and the thermal death in insects, *Nature,* 1973, 582, 1954.
4. **Church, N. S.,** Hormones and termination and reinduction of diapause in *Cephus cinctus* Nort. *(Hymenoptera: Cephidae), Can. J. Zool.,* 33, 339, 1955.
5. **Vardell, H. H. and Tilton, E. W.,** Heat sensitivity of the Indian meal moth, *Plodia interpunctella* (Hubner), immatures, *J. Ga. Entomol. Soc.,* 16, 7, 1981.
6. **Wigglesowrth, V. B.,** High temperature and arrested growth in *Rhodnius:* quantitative requirements for ecdysone, *J. Exp. Biol.,* 32, 649, 1955.
7. **O'Kasha, A. Y. K.,** Effect of sub-lethal high temperature on insect, *Rhodnius prolixus* (Stal). 1. Induction of delayed moulting and defects, *J. Exp. Biol.,* 48, 455, 1968.
8. **O'Kasha, A. Y. K.,** Effect of sub-lethal high temperature on insect *Rhodnius prolixus* (Stal). II. Mechanism of cessation and delay of moulting, *J. Exp. Biol.,* 48, 464, 1968.
9. **O'Kasha, A. Y. K.,** Effect of sub-lethal temperature on insect, *Rhodnius prolixus* (Stal). III. Metabolic changes and their bearing on the cessation and delay of moulting, *J. Exp. Biol.* 48, 475, 1968.
10. **Ivanović, J. P., Janković-Hladni, M. I., and Milanović, M. P.,** Effect of constant temperature on survival rate, neurosecretion and endocrine cells and digestive enzymes in *Morimus* larvae *(Cerambycidae: Coleoptera), Comp. Biochem. Physiol. A,* 50, 125, 1975.

11. **Ivanović, J., Janković-Hladni, M., Stanić, V., and Milanović, M.,** The role of the cerebral neurosecretory system of *Morimus funereus* larvae (Insecta) in thermal stress, *Bull. Acad. Serbe Sci. Arts,* 72(20), 91, 1980.

12. **Janković-Hladni, M., Ivanović, J., Nenadović, V., and Stanić, V.,** The selective response of the prothocerebral neurosecretory cells of the *Cerambyx cerdo* larvae to the effect of different factors, *Comp. Biochem. Physiol. A,* 74, 131, 1983.

13. **Rauschenbach, I. Y., Lukashina, N. S., Maksimovsky, L. F., and Korochkin, L. I.,** Stress-like reaction of *Drosophila* to adverse environmental factors, *J. Comp. Physiol.,* 157, 519, 1987.

14. **DeWilde, J.,** Hormones and diapause, *Proc. Int. Congr. Prog. Endocrinol.,* 184, 356, 1968.

15. **Novak, V. J. A., Mala, I., and Balaz, I.,** Effect of the prothoracic gland in *Galleria mellonella, Acta Biol. Hung.,* 25, 107, 1974.

16. **Lanzrein, B. V., Gentinetta, V., Fehr, R., and Lüscher, M.,** Correlation between haemolymph juvenile hormone titre, corpus allatum volume and corpus allatum in vivo and in vitro activity during maturation in a cockroach, *Nauphoeta cenerea, Gen. Comp. Endocrinol.,* 36, 339, 1978.

17. **Kaizer, H.,** Licht—und Elektronenmikroskopische Untersuchung der Corpora allata der Eintagsfliege *Ephemera danica* Müll. *(Ephemeroptera: Ephemeridae)* während der Metamorphose, *J. Insect Morphol. Embryol.,* 9, 395, 1980.

18. **Röseler, P. F., Röseler, I., and Strambi, A.,** The activity of corpora allata in dominant and subordinated females of the wasp *Polistes gallicus, Insectes Soc.,* 27, 97, 1980.

19. **Fluri, P., Lüscher, M., Wille, H., and Gerig, L.,** Changes in weight of pharyngeal gland and haemolymph titers of juvenile hormone, protein and vitellogenin in worker honey bees, *J. Insect Physiol.,* 28, 61, 1982.

20. **Szibbo, C. M., Rotin, D., Feyereisen, R., and Tobe, S. S.,** Synthesis and degradation of C16 juvenile hormone (JH-III) during the final two stadia of the cockroach, *Diploptera punctata, Gen. Comp. Endocrinol.,* 48, 25, 1982.

21. **Holden, J. S. and Ashburner, M.,** Patterns of puffing activity in the salivary gland chromosomes of Drosophila. IX. The salivary and prothoracic gland chromosomes of a dominant temperature sensitive lethal of *D. melanogaster, Chromosoma (Berlin),* 68, 205, 1978.

22. **Holden, J. J. A., Walker, V. K., Maroy, P., Watson, K. L., White, B. N., and Gausz, J.,** Analysis of molting and metamorphosis in the ecdysteroid-deficient mutant L(3)3^{DTS} of *Drosophila melanogaster, Dev. Genet.,* 6, 153, 1986.

23. **Fourche, J.,** Action de l'ecdyson sur les larves de *D. melanogaster* soumises au jeune. Existence d'un double conditionnement pur la formation du puparium, *C.R. Acad. Sci. Paris,* 264, 2398, 1967.

24. **Rauschenbach, I. Y., Golosheikina, L. B., Korochkina, L. S., and Korochkin, L. I.,** Genetic of esterases in *Drosophila.* V. Characteristics of the "Pupal" esterase in *D. virilis, Biochem. Genet.,* 15, 531, 1977.

25. **Rauschenbach, I. Y., Lukashina, N. S., and Korochkin, L. I.,** Role of pupal-esterase in the regulation of the hormonal status in two *D. virilis* stocks differing in response to high temperature, *Dev. Genet.,* 1, 295, 1980.

26. **Williams, C. M.,** Photoperiodism and the endocrine aspects of insect diapause, *Symp. Soc. Exp. Biol.,* 23, 285, 1969.

27. **Schneiderman, H. A. and Gilbert, L. I.,** Control of growth and development in insects, *Science,* 143, 325, 1964.

28. **Krishna, K. A. and Schneiderman, H. A.,** Prothoracotropic activity of compounds that mimic juvenile hormone, *J. Insect Physiol.,* 11, 1517, 1965.

29. **Cymborowski, B. and Stolarz, G.,** The role of juvenile hormone during larval-pupal transformation of *Spodoptera littoralis:* switchover in the sensitivity of the prothoracic gland to juvenile hormone, *J. Insect Physiol.,* 25, 939, 1979.

30. **Safranek, L., Cymborowski, B., and Williams, C. M.,** Effect of juvenile hormone on ecdysone-dependent in the tobacco hornworm, *Manduca sexta, Biol. Bull.,* 158, 248, 1980.

31. **Hiruma, K.,** Factors affecting change in sensitivity of prothoracic hormone in *Mamestra brassicae, J. Insect Physiol.,* 28, 193, 1982.

32. **Gruetzmacher, M. G., Gilbert, L. I., and Bollenbacher, W. E.,** Indirect stimulation of the prothoracic glands of *Manduca sexta* by juvenile hormone: evidence for a fat body stimulatory factor, *J. Insect Physiol.,* 30, 771, 1984.

33. **Wigglesworth, V. B.,** *Insect Hormones,* Oliver and Boyd, Edinburgh, 1970.

34. **Richards, G.,** The radioimmunoassay of ecdysteroid titers in *Drosophila, Mol. Cell Endocrinol.,* 21, 181, 1981.

35. **Richards, G.,** Insect hormones in development, *Biol. Rev.,* 56, 501, 1981.

36. **Bodenstein, D.,** Hormones and tissue competence in the development of *Drosophila, Biol. Bull.,* 84, 34, 1943.

37. **Hadorn, E.,** Differenzierungsleistungen wie derholt fragmentierter Teilstücke männlicher Genitalscheiben von *Drosophila melanogaster* nach Kultur in vivo, *Dev. Biol.,* 7, 617, 1963.

38. **Hadorn E. and Garcia-Bellido, A.**, Zur Proliferation von Drosophilazellkulturen im Adultmilieu, *Rev. Suisse Zool.*, 71, 576, 1964.
39. **Bührlen, U., Emmerich, H., and Rembold, H.**, Titer of juvenile hormone III in *Drosophila hydey* during metamorphosis determined by GC-MS-MIS, *Z. Naturforsch.*, 39, 1150, 1984.
40. **Courgeon, A. M.**, L'áctivité mitotique, en culture organotypique dans les disques oculo-antennaires de larves de *Calliphora erythrocephala* Meig., *C.R. Acad. Sci. Paris*, 268, 950, 1969.
41. **Gehring, W. J. and Nötiger, R.**, The imaginal discs of *Drosophila*, in *Developmental Systems. Insects*, Vol. 2, Counce, S. J. and Waddington, C. H., Eds., Academic Press, London, 1973, 212.
42. **Handler, A. M.**, Ecdysteroid titers during pupal and adult development in *D. melanogaster*, *Dev. Biol.*, 93, 70, 1982.
43. **Schwartz, M. B., Imberski, R. B., and Kelly, T. I.**, Analysis of metamorphosis in *Drosophila melanogaster*: characterization of giant, and ecdysteroid-deficient mutant, *Dev. Biol.*, 103, 85, 1984.
44. **Davis, K. T. and Shearn, A.**, In vitro growth of imaginal discs from *Drosophila melanogaster*, *Science*, 196, 438, 1977.
45. **Riddiford, L. M.**, Hormonal control of epidermal cell development, *Am. Zool.*, 21, 751, 1981.
46. **Sieber, R. and Bezn, G.**, Hormonal regulation of pupation in the codling moth, *Laspeyresia pomonella*, *Physiol. Entomol.*, 5, 283, 1980.
47. **Riddiford, L. M.**, Juvenile hormone in relation to the larval-pupal transformation of the cecropia silkworm, *Biol. Bull. (Woods Hole Mass.)*, 142, 310, 1972.
48. **Yagi, S.**, The role of juvenile hormone in diapause and phase variation in some lepidopterous insects, in *The Juvenile Hormones*, Gilbert, L. I., Ed., Plenum Press, New York, 1976.
49. **Hiruma, K., Shimada, H., and Yagi, S.**, Activation of the prothoracic gland by juvenile hormone and prothoracicotropic hormone in *Mamestra brassicae*, *J. Insect Physiol.*, 24, 215, 1978.
50. **Slade, M. and Zibitt, C. H.**, Metabolism of cecropia juvenile hormone in insects and mammals, in *Insect Juvenile Hormone*, Menn, J. J. and Beroza, M., Eds., Academic Press, New York, 1972, 155.
51. **Sanburg, L. L., Kramer, K. J., Kezdy, F. J., and Law, J. H.**, Juvenile hormone-specific esterases in the haemolymph of tobacco hornworm, *Manduca sexta*, *J. Insect Physiol.*, 21, 873, 1975.
52. **De Kort, C. A. D. and Granger, N. A.**, Regulation of the juvenile hormone titer, *Annu. Rev. Entomol.*, 26, 1, 1981.
53. **Shapiro, A. B., Wheelock, G. D., Hagedorn, H. H., Baker, F. C., Tsai, L. W., and Schooley, D. A.**, Juvenile hormone and juvenile hormone esterase in adult females of the mosquito *Aaedes aegypti*, *J. Insect Physiol.*, 32, 867, 1986.
54. **Rauschenbach, I. Y., Lukashina, N. S., Budker, V. G., and Korochkin, L. I.**, Genetics of esterases in *Drosophila*. IX. Characterization of the JH-esterase in *D. virilis*, *Biochem. Genet.*, 25, 687, 1987.
55. **Gilbert, L. I., Bollenbacher, W. E., Goodman, W., Smith, S. L., Agui, N., Granger, N., and Sedlak, B. J.**, Hormones controlling insect metamorphosis, *Recent Prog. Horm. Res. Proc. Laurentian Horm. Conf. (New York)*, 36, 441, 1980.
56. **Riddiford, L. M.**, Insect endocrinology: actions of hormones at the cellular level, *Annu. Rev. Physiol.*, 42, 511, 1980.
57. **Hammock, B. D.**, Regulation of juvenile hormone: degradation, in *Comprehensive Insect Physiology, Biochemistry and Pharmacology*, Vol. 7, Kerkut, G. A. and Gilbert, L. I., Eds., Pergamon Press, Oxford, 1985, 431.
58. **Nevitt, R. A. and Hammock, B. D.**, Relationship between juvenile hormone and ecdysteroid in larval-pupal development of *Trichoplusia ni (Lepidoptera: Noctuidae)*, *J. Insect Physiol.*, 32, 835, 1986.
59. **Wilson, T. G. and Gilbert, L. I.**, Metabolism of juvenile hormone I in *D. melanogaster*, *Comp. Biochem. Physiol. A*, 60, 85, 1978.
60. **Weirich, G. and Wren, J.**, Juvenile hormone esterase in insect development: a comparative study, *Physiol. Zool.*, 49, 431, 1976.
61. **Klages, G. and Emmerich, H.**, Juvenile hormone metabolism and juvenile hormone esterase titer in hemolymph and peripheral tissues of *Drosophila hydei*, *J. Comp Physiol.*, 132, 319, 1979.
62. **Hammock, B. D. and Sparks, T. C.**, A rapid assay for insect juvenile hormone esterase activity, *Anal. Biochem.*, 82, 573, 1977.
63. **Rennucci, M.**, Juvenile hormone degradation in nerve tissues and fat body of female *Acheta domesticus* (Insecta, Orthoptera), *Comp. Biochem. Physiol. A*, 84, 101, 1986.
64. **Riddiford, L. M.**, Ecdysone-induced change in cellular commitment of the epidermis of the tobacco hornworm, *Manduca sexta*, at the initiation of metamorphosis, *Gen. Comp. Endocrinol.*, 34, 438, 1978.
65. **Hwang-Hsu, K., Reddy, G., and Krishna Kumaran, A.**, Correlations between juvenile hormone esterase activity, ecdysone tier and cellular reprogramming in *Galleria mellonella*, *J. Insect Physiol.*, 25, 105, 1979.
66. **Jones, D., Jones, G., Wing, K. D., Rudnicka, M., and Hammock, B. D.**, Juvenile hormone esterase of *Lepidoptera*. I. Activity in the haemolymph during the last larval instar of 11 species, *J. Comp. Physiol.*, 148B, 1, 1982.

67. **Hodgetts, R. B., Sage, B., and O'Connor, J. D.**, Ecdysone titers during postembryonic development of *D. melanogaster*, *Dev. Biol.*, 60, 310, 1977.

68. **Garen, A., Kauver, L., and Lepesant, J. A.**, Roles of ecdysone in *Drosophila* development, *Proc. Natl. Acad. Sci. U.S.A.*, 74, 5099, 1977.

69. **Klose, W., Gateff, E., Emmerich, H., and Beikirch, H.**, Developmental studies on two ecdysone deficient mutants of *Drosophila melanogaster*, *Wilhelm Roux Arch. Dev. Biol.*, 189, 51, 1980.

70. **Borts, D. W., Bollenbacher, W. E., O'Connor, J. D., King, D. S., and Fristrom, J. W.**, Ecdysone level during metamorphosis of *Drosophila melanogaster*, *Dev. Biol.*, 39, 308, 1974.

71. **Berreur, P., Porcheron, P., Berreur-Bonnenfant, J., and Simpson, P.**, Ecdysteroid levels and pupariation in *Drosophila melanogaster*, *J. Exp. Zool.*, 210, 347, 1979.

72. **Jones, D., Jones, G., Rudnica, M., Click, A. J., and Sreekrishna, S. C.**, High resolution isoelectric focusing of juvenile hormone esterase activity from the hemolymph of *Trichoplusia ni*, *Experientia*, 42, 45, 1986.

73. **Richards, G.**, Sequential gene activation by ecdysone in politene chromosomes of *D. melanogaster*. VI. Inhibition by juvenile hormone, *Dev. Biol.*, 66, 32, 1978.

74. **Lezzi, M. and Wyss, C.**, The antagonism between juvenile hormone and ecdysone, in *The Juvenile Hormone*, Gilbert, L. I., Ed., Plenum Press, New York, 1976, 252.

75. **Kiguchi, K. and Riddiford, L. M.**, A role of juvenile hormone in pupal development of the tobacco hornworm, *Manduca sexta*, *J. Insect Physiol.*, 24, 673, 1978.

76. **Hsiao, T. H. and Hsiao, C.**, Simultaneous determination of molting and juvenile hormone titers of the greater wax moth, *J. Insect Physiol.*, 23, 89, 1977.

77. **Jones, G. and Hammock, B. D.**, Critical roles of prepupal juvenile hormone and its esterase, *Arch. Biochem. Physiol.*, 2, 397, 1985.

78. **Bhashkaran, G., Sparagana, S. P., Barrera, P., and Dahm, K. H.**, Change in corpus allatum function during metamorphosis of the tobacco hornworm, *Manduca sexta*: regulation at the terminal step in juvenile hormone biosynthesis, *Arch. Insect Biochem. Physiol.*, 3, 321, 1986.

79. **Jones, G. and Hammock, B. D.**, Prepupal regulation of juvenile hormone esterase through direct induction by juvenile hormone, *J. Insect Physiol*, 29, 471, 1983.

80. **Riddiford, L. M.**, Juvenile hormone-induced delay of metamorphosis of the viscera of the cecropia silkworm, *Biol. Bull. (Woods Hole, Mass.)*, 148, 429, 1975.

81. **Cymborowski, B.**, Effect of cooling stress on endocrine events in *Galleria mellonella*, in *Endocrinological Frontiers in Physiological Insect Ecology*, Vol. 1, Sehnal, F., Zabza, A., and Denlinger, D. L., Eds., Wroclaw Technical University Press, Wroclaw, 1988, 203.

82. **Bogus, M. I. and Sheller, K.**, Chilling stress affects the juvenile hormone synthesizing system in *Galleria mellonella* larvae, in *Endocrinological Frontiers in Physiological Insect Ecology*, Vol. 1, Sehnal, F., Zabza, A., and Denlinger, D. L., Eds., Wroclaw Technical University Press, Wroclaw, 1988, 221.

83. **Rauschenbach, I. Y., Lukashina, N. S., and Korochkin, L. I.**, Genetics of esterases in *Drosophila*. VII. The genetic control of the activity level of the JH-esterase and heat-resistance in *Drosophila virilis* under high temperature, *Biochem. Genet.*, 21, 253, 1983.

84. **Rauschenbach, I. Y., Lukashina, N. S., and Korochkin, L. l.**, Genetics of esterases in *Drosophila*. VIII. The gene controlling the activity of the JH-esterase in *D. virilis*, *Biochem. Genet.*, 22, 65, 1984.

85. **Rauschenbach, I. Y.**, Genetic control of hormone production and breakdown during metamorphosis under stress, in *Endocrinological Frontiers in Physiological Insect Ecology*, Vol. 1, Sehnal, F., Zabza, A., and Denlinger, D. L., Eds., Wroclaw Technical University Press, Wroclaw, 1988, 169.

86. **Korochkin, L. I.**, Some aspects of the gene-regulator problems in developmental genetics, *Ontogenez*, 13, 211, 1982 (in Russian).

87. **Panov, A. A.**, Response to starvation of A¹ neurosecretory cells of silkworm moth during facultative feeding, *Dokl. AN SSSR*, 176, 195, 1967 (in Russian).

88. **Applin, D. G.**, Long-term effects of diet on the neuroendocrine system of the sheep blowfly, *Lucilia sericata*, *Physiol. Entomol.*, 6, 129, 1981.

89. **Ohtaki, T.**, On the delayed pupation of the fleshfly, *Sarcophaga peregrina* Robineau-Desvoidy, *Jpn. J. Med. Sci. Biol.*, 19, 97, 1966.

90. **Ždarek, J. and Fraenkel, G.**, Overt and covert effect of endogenous and exogenous ecdysone in puparium formation of flies, *Proc. Natl. Acad. Sci. U.S.A.*, 67, 331, 1970.

91. **Rauschenbach, I. Y., Lukashina, N. S., and Korochin, L. I.**, Genetico-endocrine regulation of *Drosophila* development under extreme environmental conditions. I. A study of hormonal status and activity of JH-esterase in a *D. virilis* stock selected for resistance to a high temperature, *Genetica*, 19, 749, 1983 (in Russian).

92. **Rauschenbach, I. Y. and Lukashina, N. S.**, Genetico-endocrine regulation of *Drosophila* development under extreme environmental conditions. V. A study of hormonal status, viability and the activity of JH-esterase in *D. virilis* on a pure nutritive medium, *Genetica*, 19, 1995, 1983 (in Russian).

93. **Williams, C. M. and Adkisson, P. L.,** Physiology of insect diapause. XIV. An endocrine mechanism for the photoperiodic control of pupal diapause in the oak silkworm *Antheraea pernyi, Biol. Bull. Woods Hole, Mass.),* 128, 511, 1964.

94. **Kind, T. V.,** Neurosecretory system of some Lepidoptera in the context of diapause and metamorphosis, in *Neurosecretory Elements and Their Role in Organism,* Tyshchenko, V. P., Ed., Nauka, Moscow, 1964, 178 (in Russian).

95. **Kind, T. V.,** Possible ways of regulation of endocrine system activity of diapause, in *Photoperiodic Regulation of Seasonal Manifestations in Insects and Plants,* Tyschchenko, V. P. and Goryshina, N. I., Eds., Leningrad University Press, Leningrad, 1980, 105 (in Russian).

96. **Kind, T. V. and Vagina, N. P.,** Neurosecretory system of brain in the period of pupal diapause and metamorphosis in cutworm, *Acronycta rumicis L. (Lepidoptera: Noctuidae), Entomol. Obozr.,* 55, 286, 1976 (in Russian).

97. **Kono, Y.,** Light and electron microscopic studies on the neurosecretary control of diapause incidence in *Pieris rapae,* crucivora, *J. Insect Physiol.,* 19, 255, 1973.

98. **Yoshiaki, K.,** Endocrine activities and photoperiodic sensitivity during prediapause period in the phytophagous lady beetle *Epilachna vigintioctopunctata, Appl. Entomol. Zool.,* 15, 73, 1980.

99. **Yin, C.-M., Wang, Z. S., and Chaw, W.-D.,** Brain neurosecretory cell and ecdysiotropin activity of the non-diapausing, pre-diapausing and diapausing southwestern corn borer, *Diatraea grandiosella* Dyar, *J. Insect Physiol.,* 31, 659, 1985.

100. **Birch, L. C.,** Selection in *Drosophila pseudoobscura* in relation to crowding, *Evolution,* 9, 389, 1955.

101. **Lees, A. D.,** The physiology and biochemistry of diapause, *Annu. Rev. Entomol.,* 1, 1, 1956.

102. **Fraser, A.,** Humoral control of metamorphosis and diapause in the larvae of certain *Calliphoridae (Diptera: Cyclorapha), Proc. R. Soc. Edinburgh Sect. B,* 67, 127, 1970.

103. **Slama, K.,** Physiology of sawfly metamorphosis. II. Hormonal activity during diapause and development, *Acta Soc. Entomol. Czech.,* 61, 210, 1964.

104. **Frahm, R. R. and Kojima, K.-I.,** Comparison of selection responses on body weight under different larval density conditions in *Drosophila pseudoobscura, Genetics,* 54, 625, 1966.

105. **Fontdevila, A.,** Genotype-temperature interaction in *D. melanogaster.* I. Viability, *Genetica,* 41, 257, 1970.

106. **Mansour, M. H. and Dimetry, N. Z.,** Effect of crowding on larvae and pupae of the greasy cutworm *Agrotis ipsilon* Ufn *(Lepid.: Noctuidae), Z. Angew. Entomol.,* 72, 220, 1972.

107. **Nishigaki, J.,** Ecological studies on the cupreous chafer, *Anomala cuprea* Hope *(Coleoptera).* III. The effects of initial density of larval population on its survivorship, *Jpn. J. Appl. Entomol. Zool.,* 18, 59, 1974.

108. **Jacson, G. J., Popov, G. B., Ibrahim, A. O., Alghamedi, S. A., and Khan, A. M.,** Effect of natural food plants on the development, maturation, fecundity and phase of the desert locust, *Schistocerca gregaria* (Forskal), *Genet. Overs Res. Microsc. Rep.,* 42, 213, 1978.

109. **Torroja, E., Tome, J. M. P., and Prieto, S.,** Adaptive larval development in *Drosophila hydei* under crowded conditions, in 6th European Dros. Res. Conf., Abstract, Kupari, Yugoslavia, 1979.

110. **Arques, V. J. I. and Duarte, R. G.,** Effect of ethanol and isopropanol on the activity of alcohol dehydrogenase, viability and lifespan in *D. melanogaster* and *D. funebris, Experientia,* 36, 828, 1980.

111. **Tošić, M. and Ayala, F. J.,** Density and frequency-dependent selection at the Mdh-2 locus in *Drosophila pseudoobscura, Genetics,* 79, 679, 1981.

112. **Clark, A. D.,** Density-dependent fertility selection in experimental populations of *D. melanogaster, Genetics,* 79, 849, 1981.

113. **Mensua, J. L. and Moya, A.,** Stopped development in overcrowded cultures of *D. melanogaster, Heredity,* 51, 347, 1983.

114. **Scheiring, J. F., Davis, D. G., Ranasinghe, A., and Teare, C. A.,** Effect of larval crowding on life history parameters in *Drosophila melanogaster* Meigen *(Diptera: Drosophilidae), Ann. Entomol. Soc. Am.,* 77, 329, 1984.

115. **Berreur, P., Porcheron, P., Berreur-Bonnenfant, T., and Dray, F.,** External factors and ecdysone release in *Calliphora erythrocephala, Experientia,* 35, 1031, 1979.

116. **Jagi, S. and Fukaya, M.,** Juvenile hormone as a key factor regulating larval diapause in the rice stem borer, *Chilo suppressalis (Lepidoptera: Pyralidae), Appl. Entomol. Zool.,* 9, 247, 1974.

117. **Scheltes, P.,** Aestivation diapause in *Chilo partellus* (Swinhoe) and *Chilo orichalcociliella* (Strand), *Annu. Rep. Int. Cent. Insect Physiol. Ecol.,* 3, 210, 1976.

118. **Chippendale, G. M.,** Hormonal regulation of larval diapause, *Annu. Rev. Entomol.,* 22, 121, 1977.

119. **Chippendale, G. M. and Yin, C.-M.,** Larval diapause of the European corn borer, *Ostrinia nubilalis:* further experiments examining its hormonal control, *J. Insect Physiol.,* 25, 53, 1979.

120. **Bean, D. W. and Beck, S. D.,** The role of juvenile hormone in the larval diapause of the European corn borer, *Ostrinia nubilalis, J. Insect Physiol.,* 26, 579, 1980.

121. **Mane, S. D. and Chippendale, G. M.,** Hydrolysis of juvenile hormone in diapausing and non-diapausing larvae of the southwestern corn borer, *Diatraea grandiosella, J. Comp. Physiol.,* 144B, 205, 1981.

122. **Bogus, M. and Cymborowski, B.,** Chilled *Galleria mellonella* larvae: mechanism of supernumerary moulting, *Physiol. Entomol.,* 6, 343, 1981.

123. **McCaleb, D. C. and Kumaran, A. K.,** Control of juvenile hormone esterase activity in *Galleria mellonella* larvae, *J. Insect Physiol.,* 36, 171, 1980.

124. **Hammock, B. D., Jones, D., Jones, G., Rudnica, M., Sparks, T. S., and Wing, K. D.,** Regulation of juvenile hormone esterase in the cabbage looper, *Trichoplusia ni,* in *Regulation of Insect Development and Behavior,* Sehnal, F., Zabza, A., Menn, J. J., and Cymborowski, B., Eds., Wroclaw Technical University Press, Wroclaw, 1981, 219.

125. **Cymborowski, B., Bogus, M., Beckage, N. E., Williams, C. M., Riddiford, L. M.,** Juvenile hormone titres and metabolism during starvation-induced supernumerary larval moulting of the tobacco hornworm, *Manduca sexta L., J. Insect Physiol.,* 28, 129, 1982.

126. **Reddy, G., Hwang-Hsy, K., and Kumaran, A. K.,** Factors influencing juvenile hormone esterase activity in the wax moth, *Galleria mellonella, J. Insect Physiol.,* 25, 65, 1979.

127. **Bogus, M. I., Wisniewski, J. R., and Cymborowski, B.,** Effect of lighting conditions on endocrine events in *Galleria mellonella, J. Insect Physiol.,* 33, 355, 1987.

128. **Rauschenbach, I. Y., Lukashina, N. S., and Korochkin, L. I.,** Genetico-endocrine regulation of *Drosophila* development under extreme environmental conditions. III. The effect of crowded cultures on the viability, hormonal status and the activity of the JH-esterase in *D. virilis, Genetica,* 19, 1439, 1983.

129. **Bogus, M. I., Wisniewski, J. R., and Cymborowski, B.,** Effect of injury to the neuroendocrine system of last instar larvae of *Galleria mellonella, J. Insect Physiol.,* 32, 1011, 1986.

130. **Selye, G.,** *At the Level of a Whole Organism,* Nauka, Moscow, 1972 (in Russian).

131. **Vigas, M.,** Problem of definition of stimulus and specificity of stress response, in 3rd Symp. on Catecholamines and Other Neurotransmitters in Stress, Smolenice Castle, Czechoslovakia, 1983, 94.

132. **Sudakov, K. V.,** Specific mechanisms of emotional stress, in 3rd Symp. on Catecholamines and Other Neurotransmitters in Stress, Smolenice Castle, Czechoslovakia, 1983, 87.

133. **Singh, H. and Pandey, P. N.,** Experimental assessment of effects of larval crowding on development and reproduction in *Diacrisia obliqua* Walker *(Lep., Arctiidae), Biochem. Exp. Biol.,* 16, 157, 1980.

134. **Chernysh, S. I.,** Reaction of neuroendocrine system on injury factors, *Trudy Vses. Entomol. Ova.,* 64, 118, 1983 (in Russian).

135. **Lawrence, P. O.,** Hormonal interactions between parasitoids and hosts: adaptations to stress?, in *Endocrinological Frontiers in Physiological Insect Ecology,* Vol. 1, Sehnal, F., Zabza, A., and Denlinger, D. L., Eds., Wroclaw Technical University Press, Wroclaw, 1988, 423.

Chapter 7

THE ROLE OF STRESS IN PROCESSES OF ADAPTATION AND EVOLUTION

V. B. Sapunov

TABLE OF CONTENTS

I. INTRODUCTION

An important mechanism of biological adaptation is stress. One was described by Selye[1,2] as a generalized adaptive syndrome at the organism level. Selye considered the effect of stress in relation to mammalia. Following works showed that stress is obligatory for many groups of organisms, including insects, protista, and so on. Stress may operate not only on the organism level, but also at both tissue and cell level.[3,4] Therefore, stress is a general phenomenon for the entire living world.

Stress reaction involves three phases. The first phase is the phase of alarm, i.e., reaction to a bad condition. The second phase is increase of resistance, while the third one is super stress, or distress, leading to a decrease of the organism resistance. At the second phase viability of organisms may be increased many times. Stress therefore ensures resistance of animals in extreme and variable conditions. If it is persistance and recurrency of inappropriate states, the adaptive potencies, switched on by stress, may be exhausted. The stress reaction may be transformed into distress. One leads to suffocation of organisms. The main purpose of evolution is preservation of the population, not taking into account the fate of individuals. What about the mechanism of adaptive reaction at the population level? Let us consider the question.

II. STRESS AND ADAPTATION AT THE POPULATION LEVEL

Belyev[5] worked hard in the field of genetic results of domestication. One led, according to his data, to phenotypical variability increase. The direct cause of this is the stress effect of breeding under artificial conditions, i.e., the stress is an important mechanism acting together with natural selection. Stress may have genetic effects at both organism and population levels. Three genetic effects of stress are possible. These are

1. Switching on of some new genes. According to molecular genetic data, there are a large number of "sleeping" genes in every organism.[6] Some of the genes may be triggered by stress.
2. Change of activity of some acting genes
3. Increase of spontaneous mutability, i.e., increase of sensitivity of genes to factors inducing spontaneous mutations

The result of the process is increase of phenotypical heterogeneity at the population level. Adaptation at the level is a complicated process consisting of many stages. The onset of the process is a change of environment, and the exit is apportion of a new race. Let us trace a chain of phenomena that connect onset and exit. It is a nerve system that gives the first information on the change of ecological conditions. This system acts together with the endocrine system. By means of hormones including neurohormones stress reaction is switching on. The reaction includes many processes dealing with the organism as a whole. During stress, activity of some biochemical compounds in organisms is changed. The compounds are neurosecrets, certain ferments, and so on.[7,8] According to Selye,[1] the central mechanism of stress is the hormonal mechanism. Mammalia have corticosteroid hormones as stress regulators. Insects have ecdysone and juvenile hormones.[4,9]

Adaptation on the population level needs any specific genetic mechanisms. Hence, processes dealing with genes and gene pool are the next step of adaptation. What about a connection between this stage and the previous one? In other words, what is the mechanism of action of hormones on the genome?

TABLE 1
The Effect of Juvenile Hormone Analogs on Different Types of
Mutations, in *Drosophila melanogaster*

Variant	Type of mutation	% of mutations	n (eggs or chromosomes)
Control	Dominant lethals	1.0 ± 0.15	4267
Treatment	Dominant lethals	4.4 ± 0.48	1797
Control	Recessive semilethals in chromosome 2	0.2 ± 0.14	988
Treatment	Recessive semilethals in chromosome 2	3.5 ± 0.10	934
Control	Recessive lethals in chromosome 2	0.5 ± 0.22	988
Treatment	Recessive lethals in chromosome 2	4.8 ± 0.70	934

From Sapunov, V. B., *Dros. Inf. Serv.*, 56, 116, 1981. With permission.

III. GENETIC EFFECT OF HORMONES

The direct effect of hormones on DNA was studied in some works from 1961 by Karlson and others.[10,11] The effect may be both direct and indirect. The last one is intermediate by a chain of active chemical compounds, for example, protein-receptors of hormones. Redfield[12] showed a strong negative correlation between titer of juvenile hormone and recombination frequency in *Drosophila*. The biochemical mechanism of correlation is obscure. Shinohara et al.[13] detected that hormones of mammalia in system *in vitro* are capable of direct response to DNA. The result of the action is breaking of nucleic acid molecules. Localization of breaks is hormone-specific. There are data of the hormonal effect of puffing of polithene chromosomes of *Drosophila*.[14]

Lobashev formulated a physiological hypothesis of the mutation process.[15] According to this idea, the level of mutability depends on the state of the cells and of the organism as a whole. The status of the organism depends on the nerve and endocrine system. Hence, activity of the hormonal system has an indirect effect on the mutation process. The idea has some experimental evidence.

Serova and Kerkis[16] demonstrated the effect of mammalia hormones on chromosome aberration in the system *in vitro*. There are some data in the system *in vivo* made by this author and some others.[17-19]

In this author's previous work[17] the effect of juvenile hormone analogs was studied on strain Canton-S of *D. melanogaster*. The analogs were hydroprene and methoprene (produced by Zoecon Corporation). Microscopic examination showed dominant lethal mutation analogs induced as fertilized eggs, in which embryos had stopped development at the last stages of embryogenesis. Recessive viability mutations, induced by methoprene treatment on insects, were checked by the standard method Müller-5 (for X-chromosome) and Cy/Pm (for second chromosome), the descriptions of which are seen in the work by Medvedev.[20] Juvenile hormone analogs have an effect on the mutation frequency (Table 1). The data suggest that hormones are capable of inducing certain types of mutations. The above chemical compounds have an effect on the second chromosome, but no detectable effect on the first (that is, X) chromosome. The second one has specific regions controlling the viability of the organism of *Drosophila* as a whole.[21] At the same time such unspecific mutagens as X-rays have a less specific effect and attach all chromosomes.[22] Perhaps the endocrine system is a natural regulator of mutability in living organisms as well. Some data on the mutagenic effect of

TABLE 2

**Effect of Juvenile Hormone Analog on Morphological
Variability of Aphids (Progeny of First and Second
Generations of Treated Insects)**

		Coefficient of variation			
		Treatment of plants		Treatment of insects	
Trait	Control	F_1	F_2	F_1	F_2
Length of body	0.06	0.18	0.16	0.21	0.25
Length of antennae	0.08	0.16	0.14	0.31	0.26
Width of body	0.06	0.13	0.10	0.35	0.17

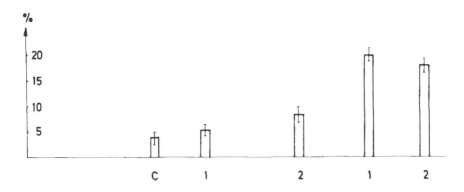

FIGURE 1.   The percent of insects with atypical color (yellow instead of wild type, i.e., green).

the juvenile hormone and its analogs are in works of L'Helias[18] and Marec et al.[19] The data are in accordance with this author's.

Genotype is a base for creating phenotype. Let us consider the phenotypical effect of hormone treatment. The effect of juvenile hormones on phenotypical variability was studied in aphis *Megoura viciae*. The methods were described in our previous work.[23,24] The variability of eight phenotypical traits was checked. This author performed data only on three of them: length of the body (out of the cauda), length of antennae with first segment, and width of body (in widest place). The traits are significant and representative, because variability of the other ones is correlated with variability of the first three.

Juvenile hormone analog (hydroprene) was applied to insect larvae, directly or over grass *Vicia faba* (sprouted beans), on which insects were bred. In both variants the variability of morphological traits within progeny of threated insects (i.e., first generation) increases (Table 2). The number of insects with atypical color increased, too (Figure 1). The effect of juvenile hormone analog took place during two generations. Hence, the effect must have a genetic basis. The heritability of traits analyzed was checked and shown to be more than (50%) the mean of the coefficient of inheritance. This fact supports the conclusion that the genetic basis of variability increase exists. The lethal effect of hormonal application existed, too, but it takes place only first generation.[24]

The data suggest that the hormones may play a role in creation of new biological forms. Hormones support natural selection by making the wide material for one.

We assumed that increase of activity of juvenile hormone leads to increase of mutation frequency. At the same time, decrease of juvenile hormone activity may have the same results. According to this author's previous data,[17] high rate of spontaneous mutability takes place in line of *D. melanogaster*, having a low titer of juvenile hormone. Data, listed above, are not contradictory. According to Nasonov and Alexandrov,[25] there are a large number of different agents which may have a similar effect on the cells. The effect consists of patterns of changes of intracellular biochemical and structural status.

The relation between some hormones is more important for intracellular status and for genetic process than absolute quantity on the same hormones. Every change of hormonal relation may prove to be the cause of the increase of mutability. Later we will consider the problem of the relation between different hormones.

IV. EFFECT OF HORMONES ON INSECT ADAPTABILITY

The mutation process is an important factor of the population's adaptation. Let us consider the effect of hormones on the adaptation at this level in aphids.

Upon change of host plant (*Pisum sativum* instead of *Vicia faba*) the phenotypical variability of parthenogenetic aphids *Megoura viciae* increased, and by the third generation fell to a control level. The treatment of insects living under favorable conditions (appropriate food) with a juvenile hormone analog (hydroprene) had about the same effect on variability. If the aphids, pretreated with the analogs, were placed in a new environment, their progeny became adapted to new conditions already in the first generation, and the variability did not exceed the control level, while the mortality exceeded 90% (in control suitable state—10%). The mortality was measured as a percentage of newborn insects that reached reproductive age.

The mechanism of hormonal effect on variability is obscure. We may propose both direct effect on genes and extra chromosomal heredity. In both cases the effect deals with inherited changes. Maybe both stress effect and hormonal treatment have a complex effect on the organism, and the general result of the last effect is phenotypical variability increase.

Summarizing, hormonal processes have an influence on the adaptation to a new environment. The stress that is a hormonal process has an effect on population adaptation, too.

V. VARIABILITY OF ORGANISMS WITHIN SUITABLE AND STRESSFUL ENVIRONMENT

There is no absolute stable and solid environment. Each ecological state is changeable, that is, it may become worse. Every new state differs from the previous one, to which the population was adapted. According to Lewontin,[26] the external conditions set certain problems that living organisms have to solve, and microevolution is a mechanism for creating these solutions.

The intrapopulation variability is a material for natural selection. The worse the external world, the more intense the natural selection. What is the rate of variability under suitable state and under stress state? The variability coefficient of quantitative traits of different species populations is relatively equal,[27] being approximately 0.1 or slightly less. This is true for many animal species living under prolonged favorable conditions. Under unappropriate conditions animal variability increases.[28] The above phenomenon was considered by quantitative analysis. This author summarized data of his own experiments on some animal species. The object of the experiments was the relation of phenotypical variability in appropriate and inappropriate conditions. The following examples of a stress situation were analyzed:

1. Change of food or host (for aphids *Megoura viciae* and *Aphis craccivora*, worm *Trichinella spiralis*)
2. Effect on insects of temperature varying from 10 to 22°C—twice per day (aphids *A. craccivora*)
3. Treatment by phosphoorganic insecticides (bug *Zygogramma suturalis*)

TABLE 3
Variability of the Length of Body under
Different Ecological Conditions

	Variation coefficient	
Species	Favorable conditions	Unfavorable conditions
Aphis craccivora	0.14	0.30
Megoura viciae	0.06	0.11
Zygogramma suturalis	0.04	0.15
Trichinella spiralis	0.07	0.21

The variability of different morphological traits of organisms was estimated. Table 3 shows variability of body length of the insects. A majority of other characters are correlated with this one. Progeny of parents, living under analyzed conditions, was measured. The data suggest that the population's variability under stress conditions increases two to three times.

What can be said about qualitative traits? Aphids *M. viciae* are green colored. Occasionally yellow insects occur. Their percentage in a stable population is about 3%; under unfavorable conditions (unusual food plant) it increases to 10%. Hence, both quantitative and qualitative variability in stressful states increase 1.5 to 3 times.

VI. MATHEMATICAL DESCRIPTION OF THE ROLE OF STRESS IN THE ADAPTATION AT THE POPULATION LEVEL

The information listed above is summarized:

1. Stress has an effect at both organisms and population levels.
2. The effect of stress on population level depends on the hormonal effect on genetic processes.
3. Under stress conditions the population variability increases.
4. The increased variability is the base of natural selection.

To assay possibilities and limits of adaptations, quantitiative estimation of the process is needed. The process using methods of applied mathematics will be described.

Characteristics of intrapopulation variability (i.e., dependence of trait frequency on its quantitative measure) may be described by normal or Gauss distribution, i.e.,

$$f(x) = \frac{1}{\sigma\sqrt{2\pi}} \exp - \frac{(x - m)^2}{2\sigma^2} \tag{1}$$

where M is mode and σ is square deviation. Under good conditions the natural selection limits variability within zone of maximal fitness (section "a" of Figure 2). The change of environment implies that the external environment sets a new problem that the living population must solve. Solution of the task means a change of traits from segment "a" to segment "b". There are two possible ways of getting a new adaptive mean. The first possibility is directional selection (Figure 3A), occurring when a new adaptive mean is relatively close to the old one. In such a situation there are a number of organisms that are capable of adaptation to a new situation, living within the old population. That is, the usual level of intrapopulation variability produces enough organisms for the founding of a new population, adaptive for a new state. An alternative situation is drawn in Figure 3B. The

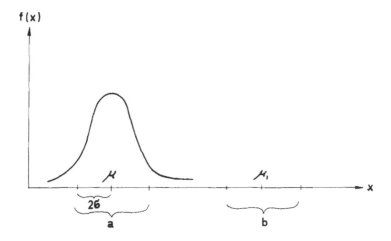

FIGURE 2. Gauss distribution of quantitative characters in population. x—measure of traits, f (x)—quota of organisms. (From Sapunov, V. B., in *Biological Indication in Anthropoecology*, Nauka, Leningrad, 1984, 195. With permission.)

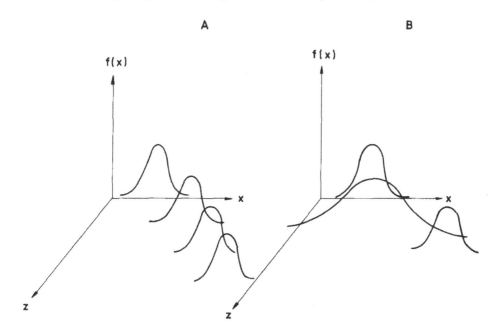

FIGURE 3. Adaptation by directional selection (A) and by genetic destabilization due to stress (B). x— measure of trait, f(x)—quota of organisms, z—generations. (From Sapunov, V. B., in *Biological Indication in Anthropoecology*, Nauka, Leningrad, 1984, 195. With permission.)

distance between "a" and "b" is longer than in the first case, indicating that the situation is extreme. Such a situation induces stress reaction. At this point we will not consider the adaptive effect of stress on the specimens. The fate of the individual is not as significant as that of the population. After stress reaction, population reproduction reduces. However, a small part of the organisms would be able to produce progeny. The genetic effect of stress is prolonged for a considerable interval of time. The genetic variability in the next generation would be increased by the hormonal mechanism, controlling variability in the parent. The increased number of variations is good material for natural selection, resulting in selection of organisms for which a new environment would prove to be suitable. If, however, despite

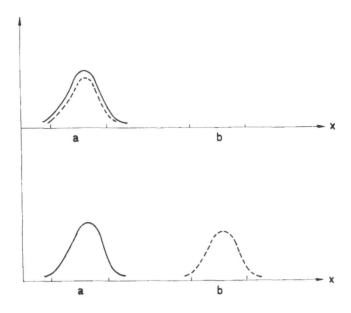

FIGURE 4.   Distribution of phenotypes (solid line) and their adaptivity (pointed line) in population. Vertical axis—function of distribution of both characters in different scales. The lines are coincident under good conditions (A); they become dispersed in changed ecological state (B). (From Sapunov, V. B., in *Biological Indication in Anthropoecology*, Nauka, Leningrad, 1984, 195. With permission.)

the genetic effects of stress and increase of variability, the organisms adapted to a new state are absent, the population would be suffocated. The quantitative aspects of the adaptation at the population level will now be described.

Under suitable conditions stabilizing selection acts in favor of organisms having maximal frequency and maximal adaptability. What does "rate of adaptability" mean? For single organisms variability may be measured as a probability to stay viable during the unit of time. For population the rate is probability to produce any number of progeny. The "rate" may be measured by the number of newborns that may reach the adult stage, per one specimen of parent generation. The adaptability (A) of stable population (i.e., population of uniform size) is

$$A = \frac{1}{(1 - My)(1 - St)} \tag{2}$$

where My is the average mortality of young animals before they reach the reproductive stage; St is percent of sterile adults. If adaptability is more than A, the population increases; if less than A the population decreases.

In stable populations distribution of adaptability is coincident with distribution of phenotype, i.e., the better the phenotype, the more frequent it is, which is the result of stabilizing selection.[29] Figure 4 shows both distributions in different scale on the vertical axis. After change of environment the adaptability of phenotypes changes, too. As can be seen, the curve of adaptability (pointed line) moves to a new position (Figure 3B). The phenotype curve (solid line) follows it by two possible ways:

1.   Directly
2.   Indirectly, through stress and destabilization (Figure 3)

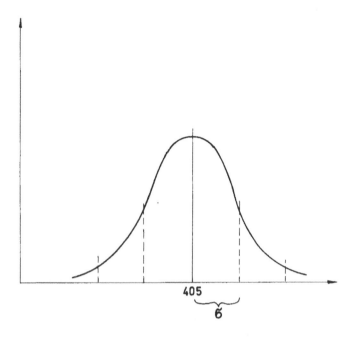

405

$\underbrace{}_{\bar{6}}$

FIGURE 5. Distribution of length of proboscis of aphids *Aphis craccivora*. On the vertical axis—the function of distribution. M = 405 mkm, σ = 35 mkm.

It must be noted that adaptability is a quantitative character. The difference between adapted and nonadapted animals is therefore relative.

As can be seen in Figure 4, the main part of the organisms living under suitable conditions are adapted to one degree or other. The percentage of nonadapted specimens is low. In checking the percentage, many biologists use the level of significant probability 0.95, i.e., 95%.[23] There is not good support of the quota. It only is established empirically. Research workers therefore usually consider variability within M ± 1.96 σ (see Figure 5). The mean 95% as a number of typical organisms in population was supported in many papers. This is the percentage of abnormality in stable human populations,[30] quota of abnormal flies in *Drosophila* populations,[31] the percent of nontypical forms in field plant populations,[32] and so on. Hence, normal organisms have a quantitative measure of analyzed traits within M ± 2 σ, which is equal to the size of segment "a" (Figure 2). However, it is quite obvious that determination of any limits is relative.

The information summarized allows for the following quantitative description of adaptive role of population stress. The area under the integral curve is equal to 1 (Figures 2 and 3). Hence, the percentage of specimens whose traits are often measured within M ± 2 σ

$$\int_{m-2\sigma}^{m+2\sigma} f(x)dx \approx 0.95 \tag{3}$$

The adaptive zone lies within "a" before change of environment, and subsequently transposes to "b" (Figure 2). Zone "a" is equal to M ± 2 σ, "b" to M_1 ± 2 σ_1. Let σ = σ_1, because the variability level in stable ecological states is the same in many populations and species.[27] As can be seen from the above, there are two pathways of adaptation, the first one being without stress reaction by directional selection (Figure 2a). The second one is the result of striking environmental change. What does striking change and little change mean?

The small distance between old and new ecological states means that number of specimens preadapted to the new state (segment ''b'') is sufficient to give rise to a new population. In parthenogenetic or hermaphrodite species the minimal number is one specimen or a little higher. For bisexual organisms the number is between 100 and 200. Let ''K'' be the number. Stressful destabilization of population means that the part of the populations preadapted to a new ecological state is too small to give rise to a new adapted population. Mathematical description of the fact is the following:

$$\int_{M_1 - 2\sigma}^{M_1 + 2\sigma} f(x)dx < \frac{K}{N} \qquad (4)$$

where N is average size of the population. The left part of the equation is a part of the population, preadapted to state $M_1 \pm 2\sigma$. Taking into account Formula (1), we may assume:

$$\int_{M_1 - 2\sigma}^{M_1 + 2\sigma} \frac{1}{\sigma\sqrt{2\pi}} \exp - \frac{(x - m)^2}{2\sigma^2} dx < \frac{K}{N} \qquad (5)$$

This is a condition of stress for average organisms in the population. Consequently, probability of stress must depend on the average and minimal sizes of the population of the species, the dependence being indirect. The step of stress reaction depends directly on physiological characteristics of the organisms. The natural selection during phylogenesis creates the characters in accordance with the average size of the population. Thus, Formulas (4) and (5) show a condition of stress for average organisms. Taking into account characteristics of Gauss distribution, one would say that 50% of the organisms are more resistant than average, and 50% are less resistant. Hence, the formulas give us the state of stress reaction for half of the population.

What is the mean value K/N? Its precise estimation is impossible in view of the wide limits of variability of population size within a species. However, we don't need exact detection. Taking into account characteristics of Gauss distribution (the rapid decrease of function after the argument has reached the value greater then 3 σ), we may assay the approximate value of K/N. The average size of the population is between hundreds and thousands. The minimal population size was considered above. Let the real value of K/N be 0.001. Using Equation 4 we will get $M_1 = M + 5.04\sigma$. Taking K/N = 0.0001 the M would be almost the same $(M + 5.5\sigma)$. Under the following decrease of K/N the M would be practically fixed. Thus, we can use K/N = 0.001 omitting another biological character of population.

There are no possible limits of change of the environment. Every change is possible and nobody can predict all the change. At the same time there exist the limits of variability of living organisms, due to fundamental characteristics of biological systems. These are physiology of the mutation process, genes interaction, etc. Consider the term ''rate of variability''. This is the limit of variation of phenotypical traits in various states. In relation to quantitative traits the following formula can be used:

$$N_\sigma = \frac{\sigma_{max}}{\sigma_{average}} \qquad (6)$$

σ_{max} is variability in extreme state; $\sigma_{average}$ is one in normal state. The essence of variability increase under an extreme situation (Figure 2) consists of the increase of possibilities for natural selection. The Formula (3) gives selection's possibility of stable population and conditions of stress reaction which occur if the possibilities are extinct. Similarly, the limits of selection after stress destabilization are described by the formula:

$$\int_{M_2 - 2\sigma_{average}}^{M_2 + 2\sigma_{average}} f(x,\sigma_{max})dx \;=\; \frac{K}{N} \tag{7}$$

If distribution is Gauss, then

$$\int_{M_2 - 2\sigma_{average}}^{M_2 + 2\sigma_{average}} \frac{1}{\sigma_{max}\sqrt{2\pi}} \exp\left(-\frac{(x - m)^2}{2\sigma^2}\right)dx \;=\; \frac{K}{N} \tag{7a}$$

When change of character "x" exceeds M_2, that is,

$$\int_{M_2 - 2\sigma_{average}}^{M_2 + 2\sigma_{average}} f(x,\sigma_{max})dx \;<\; \frac{K}{N} \tag{7b}$$

the adaptive possibilities would be exhausted. That is, new adaptive potencies, opened by stress, are fully exhausted.

What about real mean of N_σ? The similarity of variability levels of different species[27] provides the possibility to use the same constants for different species. In this author's experiments N_σ was 1.5 to 3, the mean being about 2 (see above). For a start the mean may be used. Consider examples of use of the algorithm.

Example: Adaptation of aphids to host plant

In this author's experiments the data for the example were obtained with aphids *Aphis craccivora*. The insects are parasites of plants, feeding on plant juice. Nursery insects have to pierce through cellulose walls of plants by a proboscis. The length of one is an important adaptive feature. An excessively long proboscis would not pierce through thick walls. The length of this organ therefore must be under control by natural selection. Figure 5 shows us distribution of aphids with different proboscises in a laboratory population. The mean (M) = 405 mkm; σ = 35 mkm. Let K/N = 0.001 (see above). Let natural selection of plants act against aphids by producing more thick walls. Hence, insects need to increase the length of the proboscis.

Questions are

1. What is the maximal mean of proboscis (M_1) that may be obtained by natural selection out of the destabilization?
2. What are the means that can be obtained through stress destabilization of population?
3. What is the maximal length (M_2) of the organ for this biological species?

Using Formulas 4 and 7 gives the equation:

$$\int_{M_1 - 2\sigma}^{M_1 + 2\sigma} f(x,\sigma)dx \;=\; 0.001 \tag{8a}$$

and

$$\int_{M_2 - 2\sigma}^{M_2 + 2\sigma} f(x,\sigma_{max})dx \;=\; 0.001 \tag{8b}$$

The distribution, performed in Figure 5, is Gauss. Hence, use Formulas (4) and (8). Let N_σ = 2. The solutions of the above equations are

$$\frac{1}{35\sqrt{2\pi}} \int_{M_1 - 70}^{M_1 + 70} \exp - \frac{(x - 405)^2}{2.35^2} \, dx \;=\; 10^{-3}$$

and

$$\frac{1}{70\sqrt{2\pi}} \int_{M_2-70}^{M_2+70} \exp - \frac{(x - 405^2)}{2.70^2} \, dx = 10^{-3}$$

Computation allows for the use of a general solution. Results are

$$M = M + 5.04\sigma \qquad (9)$$

$$M_2 = M + 8.14\sigma \qquad (10)$$

Finally, results on the base of Formulas (9) and (10) are

$$M_1 = 580 \text{ mkm}$$
$$M_2 = 685 \text{ mkm}$$

The answers to questions of the task are

1. Directional selection may direct the length of proboscis up to 580 mkm.
2. If, however, insects need a longer proboscis than that of the length up to 580 mkm, the population comes under stress.
3. The length exceeding 685 mkm is impossible for the species. There are no potencies for a continuous rise of proboscis in the gene pool of the population. Insects are unable to inhabit plants in which cell walls are thicker.

We assume that under stable states distribution of phenotypes and that of their fitness are coincidental (see Figure 4). We may therefore use the same mathematical model for study of adaptive limits of organisms.

We have considered the role of the stress in adaptation and microevolution in relation to quantitative traits that are distributed in the population according to Gauss law. What about the traits, distributed by another low? Asymmetrical distribution may be described by dependence e^{-x^n}, $1/x$, and so on. Figure 6 shows distribution within a laboratory population of fly *D. melanogaster,* females per quota of undeveloped eggs. The distribution is Poisson.

Occasionally asymmetrical distribution can be extrapolated by Gauss distribution, taking into account only one branch (Figure 7). Such a type of distribution deals with sensitivity of organisms to nocuous agents. Distribution of an optimal dose of toxic factor is accorded to Figure 7, that is, this dose for the majority of organisms is zero. The higher the dose, the smaller is the number of organisms for which the dose is optimal. This pattern of distribution may be used in solving certain tasks in relation to analysis of a critical dose of nocuous compounds for organisms and for population.[33] The mode of computing practical tasks is the same that was considered above.

In summary, it must be stressed that mathematical description is a relative approximation to nature. There exist two practical types of distribution: symmetrical and asymmetrical. These can be extrapolated by graphs from Figures 2 and 7. Such graphs may be analyzed by methods of applied mathematics.

VII. THE STRESS AS EXAMPLE OF GENERALIZATION OF BIOLOGICAL REACTIONS

We have assumed that term ''stress'' stands for a number of biological phenomena. Stress occurs at some structural levels, namely, at tissue, organism, and population levels.

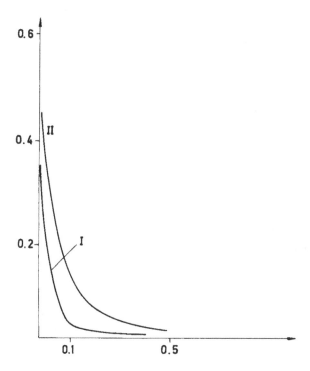

FIGURE 6. Distribution of females of *Drosophila melanogaster* by the number of suffocated eggs. I—late embrionic lethals, II—early lethals. The horizontal axis—number of dead eggs; the vertical axis—number of females.

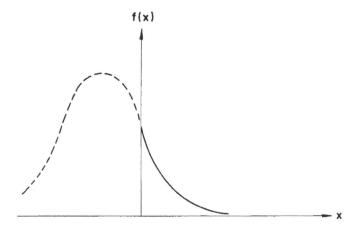

FIGURE 7. Extrapolation of asymmetrical distribution by Gauss distribution.

Stress is a variant of responding of the organism to environmental stimuli. The number of extreme factors is unlimited. Living systems are unable to predict all tasks that nature may impose on them. It stands to reason, therefore, that evolution created certain universal reactions to every action by the environment. The tendency of the existence of such a reaction is generalization of biological reactions. It is an important principle of living system action.

Stress appeared to be a most notable example of the principle. Selye[1] considered stress as a generalized adaptive syndrome at organism level. To further develop the idea, this author has considered the effect of stress at the population level. The universal mode of adaption on the last level is a variability increase. A wide spectrum of forms is essential material for selection of a new population, adapted to a new ecological state.

Apart from the organism level, Selye[1] considered the possibility of a generalized reaction on the tissue level. The reaction on the cellular level has certain specific features depending on structure of cells and characteristics of effectors. At the same time there are some common characters of reaction of different cells. Nasonov and Alexandrov[25] described the pattern of uniform reactions of different cells as "paranecrosis". One may be considered as a stress (or something like stress) at the cell level.

Changes on ontogenesis are controlled by the environment. Reactions on external actions are generalized, too. For example, phenotype "abnormal abdomen" is well known in fly *Drosophila melanogaster*. It is due to the destruction of abdominal sclerotization. The phenotype may be induced by many different agents influencing morphogenesis. The agents are change of temperature, deficiency as well as surplus of juvenile hormone and of ecdysone, unspecific protein's treatment, etc.[34-36] In other words, apportion of the abnormal phenotype is a certain morphogenic "stress". Hence, the generalized reaction of the level of morphogenesis exists.

An important task of insect endocrinology is the relation between two main hormones, i.e., ecdysone and juvenile hormone, whose interaction remains obscure. There exist three possibilities: antagonistic, synergistic, and permissive.[37] Within any physiological limits hormones act as antagonists, and beyond the limits they act in the single direction. According to the generalization principle, there seems to be no contradiction in the data. Every hormone has its own physiological effect. The superactivity of hormones triggers the generalized reaction of cellular receptors. Hence, an important feature of living systems is the possibility for generalized reactions of external effectors. Generalization means that a limited number of possible reactions are accorded to an unlimited number of effectors.

Generalization occurs at some structural level. We have summarized generalization on cellular, tissue, organism, and population levels. Presumably, that generalization on other structural levels exists.

VIII. CONCLUSION

Stress was considered by Selye[1,2] as a generalized adaptive syndrome at the organism level. The idea was supported by experiments with mammalia. Further works suggested that stress is a phenomenon common to all living worlds, and can act at different structural levels. Hormones provide the biochemical basis of stress. In mammalia they are steroid ones. In insects ecdysone and juvenile hormone act as regulators of stress reactions. As is known, during stress the hormonal status of an organism is changed. The hormones control some genetic processes. Change of endocrine status may have three effects on the genetic system:

1. Increase of speed of spontaneous mutation apportions
2. Switching on of some "sleeping" genes
3. Change of gene activity

Stress, therefore, appears to have two results: the first one is increase of organism resistance to nocuous agents; the second is increase of variability of progeny. At the onset of stress reaction, reproductive potencies are decreased, but are restored after some time. The genetic effect of stress provokes to be prolonged for more time. That's why changes of genome are transmitted to following generations. Results of the transmission are an increase of variability at the population level, a factor enlarging possibility for natural selection. Stress thus intensifies possibilities for microevolution. There are two types of inherited variability of population, both open and cryptic. The open one is visible in a stable population living under good ecological conditions. If the variability is not sufficient for adaptation to a new environment, the stress reaction begins. The stress opens the second, that is, cryptic variability. One is encoded by genetic pool of population. This form of variability serves as material for adaptation to an extreme status of ecological factors. The stress reaction makes internal variability actual and visible. Both visible and invisible variability have limits determined by a species gene pool. The limits may be assayed on the base of a study of phenotypical variability by methods of population genetics and applied mathematics. The knowledge of the limits is of practical importance. Stress is an example of principle of biological reaction's generalization. Hence the principle is that living systems have a limited number of possible reactions for an unlimited number of environmental effectors.

The main objective of modern biology is the understanding of evolution. Adaptive microevolution is a part of one. To understand the modes of evolution an integrated approach is needed using knowledge from different fields of biology, such as physiology, ecology, biochemistry, etc.

IX. SUMMARY

Selye[1,2] considered stress as a generalized adaptive syndrome at the organism level. Later research suggested that stress has an effect not only on the organism adaptive potencies. According to Belyev,[5] stress is a factor of evolution. It is this idea that underlies the present paper. An important feature of stress reaction is the role played by hormones. Unsuitable ecological states act on the endocrine system of organisms which, in turn, act on the hormonal status within the organism. According to our own data and those available in literature, hormonal disturbance may affect genetic processes, this leading to genetic and phenotypical variability increase. That may also result from extreme ecological states. Stress reaction therefore appears to have two results, the first one being increase of adaptability of specimens, the second increase of variability within a population, whose wide spectrum of forms provides sample material for natural selection that selects organisms adapted to a new environment. It stands to reason, therefore, that stress reaction is a factor of adaptation at the population level. Cryptic variability, encoded in the genetic system of population, serves as material for adaptation. Stress reaction makes internal variability actual.

Both cryptic and actual variabilities have limits determined by species gene pool. These limits are analyzed in the paper using population genetics and applied mathematics. Knowledge of the limits is of great practical importance for the solution of many practical tasks.

The biological reaction on environment affectors is generalized, i.e., a limited number of reactions are accorded to an unlimited number of affectors. Generalization of biological reactions is a fundamental characteristic of living systems. This character was created during evolution.

Stress is one of a certain generalized reaction.

REFERENCES

1. **Selye, H.**, Syndrome produced by diverse nocuous agents, *Nature*, 138, 32, 1936.
2. **Selye, H.**, *Stress in Health and Disease*, Butterworths, London, 1976.
3. **Chaikovsky, S. V.**, Natural selection as processing of genetical information, *Theor. Exp. Biophys. (Kaliningrad)*, 6, 148, 1976 (in Russian).
4. **Chernysh, A. I.**, Reaction of neuroendocrine system on nocuous agents, *Ann. All Union Entomol. Soc.*, 64, 118, 1983 (in Russian).

5. **Belyev, D. K.,** Destabilizing selection as a factor in domestication, *J. Hered.,* 70, 301, 1979.

6. **Zuckerkandl, E. and Pauling, L.,** Evolutionary divergence and convergence in protein, in *Evolving Genes and Proteins,* Academic Press, New York, 1976, 270.

7. **Ivanović, J., Janković-Hladni, M., Stanić, V., and Milanović, M.,** The role of cerebral neurosecretory system of *Morimus funereus* larvae (Insects) in thermal stress, *Bull. 72nd Serbe Acad. Sci. Nat.,* 20, 91, 1980.

8. **Janković-Hladni, M., Ivanović, J., Stanić, V., Milanović, M., and Božidarac, M.,** Possible role of neurohormones in the process of acclimation and stress in *Tenebrio molitor* and *Morimus funereus.* Changes in the activity of digestive enzymes, *Comp. Biochem. Physiol. A,* 67, 477, 1980.

9. **Sapunov, V. B.,** On the role of endocrine system in the process of appearance of mutations, *J. Obshei Biol.,* 41, 192, 1980 (in Russian).

10. **Karlson, P.,** Regulation of gene activity by hormones, *Humangenetics,* 6, 99, 1968.

11. **Zalokar, M.,** Effect of corpora allata on protein and RNA synthesis in colletarial glands of *Blatella germanica, J. Insect Physiol.,* 14, 1177, 1968.

12. **Redfield, H.,** Delayed mating and the relationship of recombination to maternal age of *Drosophila melanogaster, Genetics,* 53, 593, 1966.

13. **Shinohara, K., Hiroshisa, O., Yoshida, T., and Kazuo, Y.,** Mode of actions of steroid hormones, catecholamines and hexose oximes on DNA, *J. Fac. Agric. Kuyshi Univ.,* 19, 169, 1975.

14. **Ashbürner, M. and Richards, G.,** The role of ecdysone in the control of gene activity in the polythene chromosomes of *Drosophila, in Insect Development,* Oxford University Press, New York, 1976, 203.

15. **Lobashev, M. E.,** Physiological (paranecrotical) hypothesis of mutation process, *Vestn. Leningr. Univ.,* 8, 10, 1947 (in Russian).

16. **Serova, I. A. and Kerkis, Y. J.,** Cytogenetical effect of some steroid hormones. The study of mutagenic activity of hystamine and bradikinin in culture of cells of mammalia and man, *Dokl. Acad. Sci. U.S.S.R.,* 232, 478, 1977 (in Russian).

17. **Sapunov, V. B.,** The effect of juvenile hormone analogs on mutation frequency in *Drosophila melanogaster, Dros. Inf. Serv.,* 56, 116, 1981.

18. **L'Helias, C. and Proust, J.,** Mutation induced by a hormonal imbalance in *Drosophila melanogaster, Mutat. Res.,* 93, 125, 1982.

19. **Marec, F., Socha, R., and Gelbič, I.,** Mutagenicity testing of the juveniod methoprene (ZR-515) by means of *Drosophila* wing spot test, *Mutat. Res.,* 188, 209, 1987.

20. **Medvedev, N. N.,** *Practical Genetics,* Nauka, Moscow, 1968 (in Russian).

21. **Pole, I. R.,** Analysis of Genetical Determination of Sexual Activity in *Drosophila melanogaster,* Ph.D. thesis, Leningrad, 1979 (in Russian).

22. **Dubinin, N. P.,** Evolution of populations and radiation, *Atomisdat, (Moscow),* p. 1, 1966 (in Russian).

23. **Sapunov, V. B.,** Population stress as biological indicator of ecological disturbances, in *Biological Indication in Anthropoecology,* Nauka, Leningrad, 1984, 195 (in Russian).

24. **Sapunov, V. B. and Kuznetsova, V. G.,** The use of aphids for genetical investigations: the state and the perspectives; population structure, genetics and taxonomy of aphids and thysomoptera, *Ac. Publ. (Hague),* p. 139, 1987.

25. **Nasonov, D. N. and Alexandrov, V. J.,** *Reaction of Living Substance on Extremal Effectors,* Nauka, Moscow, 1940 (in Russian).

26. **Lewontin, R.,** Adaptation, *Sci. Am.,* 239, 156, 1978.

27. **Cherepanov, V. V.,** *Evolutionary Changeability of Land and Water Animals,* Nauka, Novosibirsk, 1986 (in Russian).

28. **Parsons, P. A.,** Evolutionary rates under environmental stress, *Evol. Biol.,* 21, 311, 1987.

29. **Bochkov, N. P., Ivanov, V. I., and Zacharov, A. F.,** *Medical Genetics,* Meditsina, Moscow, 1985 (in Russian).

30. **Shmalgausen, I. I.,** *Factors of Evolution,* Nauka, Moscow, 1968 (in Russian).

31. **Dodzhansky, Th.,** *Genetics of the Evolutionary Process,* Columbia University Press, New York, 1970.

32. **Krenke, N. P.,** Somatic traits and factors in apporation of forms in, *Phenigenetical Variability,* Vol. 1, Nauka, Moscow, 1933, 11.

33. **Sapunov, V. B. and Karelin, O. A.,** Approximate estimation of undangerous level of chemical compounds in air on the base of principles of population genetics, *Hyg. Sanit.,* 5, 15, 1986.

34. **Hillman, R., Shafer, S. Y., and Sang, J. H.,** The effect of inhibition of protein synthesis on the phenotype abnormal abdomen, *Genet. Res.,* 21, 229, 1973.

35. **Postlethwait, J. H.,** Juvenile hormone and adult development of *Drosophila, Biol Bull.,* 147, 119, 1974.

36. **Sapunov, V. B.,** Generalization of living system's reactions, in *Problems of Analysis of Biological Systems,* Moscow University Publishers, Moscow, 1983, 120.

37. **Willis, J. H. and Hollowell, M. P.,** The interaction of juvenile hormone and ecdysone-beta: antagonism, synergystics or permissive?, in *The Juvenile Hormones,* Academic Press, New York, 1976, 270.

Chapter 8

THE PRACTICAL IMPORTANCE OF STRESS STUDIES

Miroslava I. Janković-Hladni and Jelisaveta P. Ivanović

TABLE OF CONTENTS

I. INTRODUCTION

It is apparent that contemporary research no longer recognizes the artificial border line of applied vs. basic research. For the joint progress of both investigations one of the most important aspects is the collaboration between the scientists involved in these investigations, in terms of exchange of ideas, results, experience, and sharing of cooperative programs.

The study of stress in insects represents ideal opportunity for the feedback between basic and applied investigations. The first investigations concerning stress in insects resulted from the necessity to control some insect species, and were carried out within the studies concerning the action and mechanism of action of insecticides applied to control harmful insects.

Results on nonspecific response of insects to insecticides and other stressors at different levels of biological organization (organism, the nervous and neuroendocrine systems, metabolism) confirmed Selye's hypothesis that in insects also general adaptation syndrome plays a role. These responses represent the baseline of these investigations (see Chapters 1 and 2).

Regardless of the fact that a number of fundamental problems concerning stress in insects is still unsolved, the question arises of whether it might be expected that in the near future the available results on stress would have practical significance.

II. THE INTERRELATIONSHIP BETWEEN STUDIES ON STRESS AND INSECT CONTROL

We shall try to suggest some of numerous possible applications of the results obtained, concerning the stress in insects, in the control of insect pests.

It should be emphasized, first of all, that the population density of one harmful species of insects may be successfully controlled only if it is thoroughly studied from different aspects: ecological, ecophysiological, physiological, biochemical, and genetic. The studies, directed to monitoring of the response and mechanism(s) of the response of insects to stressors, are of particular interest for the applied investigations. It should be mentioned that some exceptionally harmful species were extensively studied from some aspects and poorly, if at all, from others. Thus, for example, gypsy moth *Lymantria dispar* (Lepidoptera), which is a defoliator of forest and fruit trees, is one of the most dominant pests in the U.S. It is thoroughly studied from ecological and genetic standpoints, but physiological and biochemical aspects were poorly approached. More recently, considerable interest has been expressed in the latter aspects of this species and relevant investigations have been started.

It is well known that insecticide treatment as well as that of other pesticides is a growing danger to all living organisms, including humans, because it causes environmental pollution. It is also known that it is costly to obtain currently used and new insecticides (synthesis, testing, production, application). Besides, development of resistance to insecticides in treated insects complicates the matter.

In applying a wide variety of insecticides (1st, 2nd, 3rd, 4th, and ...n generation) one must primarily bear in mind environmental protection on one hand and economy on the other hand.

From the aspect of stress, the motto "the right dose at the right time" is ever more meaningful. This motto becomes a must in the control of pest insect species. In studying the stress in insects at different levels of biological organization (population, organism, the nervous and neuroendocrine systems, cell, molecule) it will be possible to define more precisely the right dose and the right time of application of adequate chemical or biological agents.

III. THE LETHAL DOSE DEPENDENCE ON THE COMBINED EFFECT OF INSECTICIDES AND OTHER STRESSORS

Insecticides applied to natural populations exert their influence in combination with other ecological factors (temperature, humidity, food, etc.). Thus, dependent on the combination of ecological factors of various intensity, the range of insecticide lethal dose must be determined, which would enable their more rational use. It should be noted that the effect of pyretroids in *Musca domestica* is greater if environmental temperature is 18°C than if it is 32°C.[1] In contrast, in the three species of *Lepidoptera* the linear correlation between temperature and pyretroid toxicity is observed.[2]

It is well known that modern insecticides act by disruption of the nervous and neuroendocrine system functions, which is also the case with other stressors (e.g., temperature, starvation, physical stress). Combined effect of stressory temperature and other stressory factors on the protocerebral neurosecretory cells and metabolism of *M. funereus* larvae induces significantly different responses at the level of neurosecretory cells and metabolism. Namely, disturbances of hormonal balance were of different degrees.[3,4] Relevant investigations, especially those regarding testing of hormone(s) and neurohormone(s) as possible insecticides, would be not only of fundamental but also of applicable significance. In the last few years a considerable progress was made in this field—several neurohormones were synthesized and identified—thus becoming candidates for insect pest control.[5]

IV. THE PHYSIOLOGICAL STATE OF THE INSECT AND THE RIGHT TIME OF INSECTICIDE APPLICATION

The dose of insecticide and timing of its application are also dependent on the physiological state of the pest to be treated (e.g., developmental phase, phase of circadian or annual rhythm). Physiological state is determined by many factors (genetic, ecological) and actually reflects hormonal and metabolic status of the insect studied during its life cycle. It is well known that during the process of development the hormonal balance is changed. The transition from one hormonal level to another represents critical phases, i.e., the periods of the increased susceptibility of insects to the effects of environmental factors, including insecticides. In many insect species critical phases were differentiated during the development (e.g., prior to and upon moulting, younger stages in comparison to older ones, etc.). Data regarding changes in hormonal level during the circadian or annual cycle are less numerous. Still, critical phases were also suggested in some species of insects.[6] It seems of particular importance to determine precisely these critical phases in the pests to be treated. Namely, it might be suggested that the insecticide dose to be applied during the critical phase (e.g., after moulting, dawn-dusk, spring-autumn) is lower than that to be applied during the stable phase (e.g., intermoult, day-night, summer-winter).

During the evolution in many species, as a response to unfavorable factors, diapause arises. This is a physiological state in the process of development in which diapausing insects are resistant to the effect of environmental factors, including insecticides. There is relatively extensive literature on diapause from ecological, ecophysiological, and physiological aspects. However, in insects the onset and termination of diapause are also critical phases during which hormonal balance is disturbed. As lower insecticide doses are lethal during the onset and termination of diapause, it might be recommended that pests should be treated during these phases.[7]

V. CONCLUSION

At the end of this short outline of possible applications of the results obtained in studying stress in insects, it may be concluded that in the near future the results of these and further investigations concerning stress will be among the indispensable components of the integrated Insect Pest Control Programme.

REFERENCES

1. **Scott, G. J. and Georghian, G.,** Influence of temperature on knockdown toxicity, and resistance to pyrethroids in the house fly, *Musca domestica, Pest. Biochem. Physiol.,* 21, 531, 1984.
2. **Sparks, T. C., Shour, M. H., and Wellemeyer, E. G.,** Temperature-toxicity relationships of pyrethroids on three Lepidopterans, *J. Econ. Entomol.,* 75, 643, 1982.
3. **Ivanović, J., Janković-Hladni, M., Stanić, V., and Kalafatić, D.,** Differences in the sensitivity of protocerebral neurosecretory cells arising from the effect of different stressors in *Morimus funereus* larvae, *Comp. Biochem. Physiol.,* 80A, 107, 1985.
4. **Janković-Hladni, M., Ivanović, J., Stanić, V., and Milanović, M.,** Effect of different factors on metabolism of *Morimus funereus* larvae, *Comp. Biochem. Physiol.,* 77A, 351, 1984.
5. **Keeley, L. L.,** Speculations on biotechnology applications for insect neuroendocrine research, *Insect Biochem.,* 17, 639, 1987.
6. **Ivanović, J., Janković-Hladni, M., Stanić, V., Milanović, M., and Nenadović, V.,** Possible role of neurohormones in the process of acclimatization and acclimation in *Morimus funereus* larvae *(Insecta)*. I. Changes in the neuroendocrine system and target organs (midgut, haemolymph) during the annual cycle, *Comp. Biochem. Physiol.,* 63A, 95, 1979.
7. **Denlinger, D. L., Giebulowitz, J., and Adeokun, T.,** Insect diapause: dynamics of hormone sensitivity and vulnerability to environmental stress, in *Endocrinological Frontiers in Physiological Insect Ecology,* Sehnal, F., Zabza, A., and Denlinger, D. L., Eds., Technical University Press of Wroclaw, Wroclaw, Poland, 1988.

Chapter 9

SUMMARY AND PERSPECTIVES OF FUTURE STUDIES

Jelisaveta P. Ivanović and Miroslava I. Janković-Hladni

In the last few years a growing number of scientists have indicated the universal phenomenon of stress. Namely, they suggest that stress exists not only in mammals and vertebrates but also in organisms whose organization is by far less complex, i.e., in representatives of invertebrates and even in plants. Among invertebrates, insects and sea mollusks were most extensively studied from that aspect.

There are several definitions of stress but they can be reduced to the following two: The first defines stress as the state of the organism, caused by the stressor or by stressors, which always leads to the organism's death. According to the second definition, stress reflects the state of the organism elicited by the effect of one or several stressors which does not result in the animal's death. Due to the reorganization of metabolism, which is either intensified or depressed in respect to the initial level, changes in the developmental rate occur, being either slowed down or accelerated, resulting in the termination of the development or the death of the organism. Both definitions are true if stress is considered at the level of population to which the organism belongs.

Population of any organic species possesses physiological and biological plasticity, which includes not only the response to stressor but also relates to the type of the response. The plasticity of the population reflects the sum of plasticities of individuals of which it is composed (Σip) and thus plasticity of individuals and mortality are inversely proportional. Plasticity is gene controlled. Hence, there are differences between the response to stress of the organism and that of the population. However, in studying the stress in the organism the interrelationship between organism and population should not be neglected. During the larval development, nutritional and thermal stresses frequently affect fertility in imago of both sexes, sex ratio, etc. which results in the change of population density, one of the most essential attributes of a population.

Under the effect of various stressors the insects employ several strategies but the most common are: synthesis of protective substances and isoenzymes, compensatory reactions at the level of metabolism, functional enzymes, modulation of membranes, and others. Thus, for example, during thermal stress, i.e., during heat stress, the heat shock proteins (hsp) which have a protective role are synthesized. Under the effect of near zero and subzero temperatures, cryoprotective, either carbohydrate (polyols, sugar) or proteinaceous substances - ice nucleating agents (INA) and thermal hysteresis proteins (THP) are synthesized. Frequently, some free amino acids (proline, alanine) are also synthesized in response to these temperatures. Compensatory reactions are especially pronounced in functional proteins of intermediary metabolism and of the digestive system which play a major role in energy metabolism. Energy metabolism is of crucial significance for the outcome of stress since its influence on the adaptive capacity is decisive. Change in the pH of extracellular fluids, tissues, and cells should not be disregarded because it affects the type of the response of cellular membranes, conductance of the nervous system, and the number and function of the variety of channels.

It should be pointed out that phytophagous insects in the course of evolution of the interrelationship insect-plant began producing nonspecific functional proteins (complex of microsomal oxidases), which are used primarily as protective enzymes against plant toxicants and secondarily against thermal and chemical stressors.

As judged by recent findings, the initial link in the response of insects to stress is biogenic amines, most frequently octopamine, which is indicative of a certain similarity to the initial phase of stress in vertebrates. In the later phase of stress neurohormones, released and later also synthesized, substituted for biogenic amines. Neurohormones, along with other hormones (ecdysteroids, JH) accelerate or slow down the development. Shortening of the development may lead to changes in the individual's biological potential, whereas prolonged development very often results in the animal's death due to its inability to pass to another stage, i.e., to metamorphose.

In the future, investigations concerning the stress, which might be said to be still in the pioneering phase, should result in the proposal of the model of stress in insects. To achieve the goal much has to be done in that respect. Primarily, the transition of a certain environmental factor to the stressor should be quantitatively expressed, which is the prerequisite for the objective definition of the stress as the biological phenomenon.

Besides, in studying the phenomenon of stress both in insects and other organisms, it must be taken into account that the organism is a biological unit which is the integral part of evolutive unit-population. One of the key problems to be solved, prior to the establishment of the model of stress, is the problem of the interrelationship organism-population, but approached multidisciplinarily from physiological, genetic, ecological, and evolutive aspects.

In addition, regardless of the fact that the organism responds to stressor as a whole, it should not be disregarded that this general response is composed of the responses at different levels of biological organization. In this respect the response of insects to the effect of stressors at cellular and subcellular levels is least studied. In order to find out more exact explanations for the specificities and nonspecificities of certain types of responses to stressors it is necessary to intensify the research on stress from the aspect of molecular biology in order to recognize the role of the population of receptors, mitochondria, and membranes in insect stress.

Though in the last few years attention has been given to the study of the role of biogenic amines in stress in insects, still endocrinological research concerning stress in insects is scarce. It is necessary to obtain insight into the interrelationship between the action of biogenic amines, observed during the first phase of stress, and neurohormone release and synthesis. Stress in insects influences the developmental processes by changing the developmental rate, the consequence of which is either the shortening or prolongation of the development. Change in the developmental rate is interrelated with the activation of ecdysteroids and JH. However, mechanism of the selective triggering of either of the two hormones is unknown yet.

Designing of the model of stress for vertebrates, including that for mammals, as well as establishment of the model of stress for invertebrates, including that for insects, are indispensable prerequisites for the proposal of the general model of stress in all living beings.

INDEX

H

I

Printed and bound by CPI Group (UK) Ltd, Croydon, CR0 4YY

22/10/2024

01777600-0017